THE QUIET ELEGANCE OF NUMBERS

MATHEMATICS AND THE GENTLE MACHINERY OF THE UNIVERSE

DR ALOKE MULLICK

Made with ♥ on the Notion Press Platform
www.notionpress.com

Dedicated to

My Parents –

Dr Lalita Mullick & Er ML Mullick, for teaching me Life's Values.

My Spiritual Masters-

Babuji, Chariji and Daaji, for showing me Life's True Direction.

My Family –

Dr Charu Mullick, Kartikey and Swarna, for loving me despite all my imperfections.

Contents

Preface *vii*

Prologue *xi*

1. The Most Unreasonable Success Of Mathematics 1
2. Counting, Curiosity And The Dawn Of Numbers 7
3. Geometry: The Universe Drawn With A Straightedge 14
4. Algebra: When Numbers Learn To Converse 20
5. Infinity, The Number That Refuses To Behave 26
6. Calculus And The Art Of Change 32
7. Mathematics And The Music Of The Universe 39
8. Probability: When Mathematics Learns To Gamble 46
9. Mathematics And The Architecture Of Physics 52
10. Patterns Of Life: Mathematics In Biology 59
11. Computers: When Mathematics Learned To Think Quickly 66
12. Srinivasa Ramanujan: The Poet Of Numbers 73
13. Other Indian Luminaries Of Mathematics 80
14. Elegance: Why Mathematicians Fall In Love With Equations 86
15. The Mind Of Mathematics: Discovery, Invention, And Intuition 93
16. Why The Universe Speaks Mathematics 100
17. Does God Love Mathematics? 106
18. The Future Of Mathematics And The Endless Adventure 112

Epilogue 119

Preface

There is a particular kind of hesitation that often accompanies the mention of mathematics.

It is not quite fear, though it occasionally borrows its tone. Nor is it indifference, though it sometimes disguises itself as such. It is, perhaps, a quiet suspicion that mathematics is a subject best left to those who possess an unusual fondness for symbols, a tolerance for abstraction, and a willingness to spend long hours contemplating matters that appear, at first glance, to have little to do with everyday life.

This book is written in gentle disagreement with that suspicion.

Mathematics, despite its reputation, is not an isolated discipline. It is not confined to blackboards, examinations, or the more determined corners of academia. It is, in fact, woven into the fabric of the world in ways that are both subtle and surprisingly pervasive.

It appears in the patterns of nature, in the rhythms of music, in the structure of physical laws, and even in the uncertainties of chance. It governs the motion of planets and the growth of populations, the flow of electricity and the spread of information. It is, quite simply, one of the most effective ways we have of understanding the universe.

And yet, it rarely presents itself with fanfare.

Mathematics does not insist on being noticed. It operates quietly, often behind the scenes, providing structure without demanding attention. One may go about one's daily life—checking the time, navigating a route, enjoying a piece of music—without consciously engaging with the mathematics that makes these activities possible.

This unobtrusive nature is both a strength and a disadvantage.

On the one hand, it allows mathematics to function seamlessly within the world. On the other, it means that its presence is easily overlooked. When it does come into focus, it is often in the form of formulas and procedures, presented without context or narrative, which can make it appear more forbidding than it truly is.

The intention of this book is to present mathematics in a different light.

Rather than approaching it as a collection of techniques to be mastered, we will explore it as a story—one that unfolds across time, cultures, and disciplines. It is a story of curiosity, of patterns noticed and pursued, of ideas developed with care and, occasionally, with a certain amount of stubbornness.

It is also a story of connection.

Each chapter in this book examines a particular aspect of mathematics and its relationship with the world. We begin with the curious observation that mathematics seems to describe the universe with remarkable accuracy. We then trace its development from simple counting to more sophisticated concepts such as geometry, algebra, and calculus.

Along the way, we encounter infinity—an idea that refuses to behave according to ordinary expectations—and probability, which attempts to make sense of uncertainty.

We then explore how mathematics interacts with the sciences: how it provides the framework for physics, the patterns underlying biology, and even the structure of music.

These connections are not incidental.

They suggest that mathematics is not merely a human invention, but a language that resonates with the underlying structure of reality. Whether one believes that mathematics is discovered or created, its effectiveness is undeniable.

At the same time, this book does not seek to overwhelm the reader with technical detail.

The emphasis is on understanding rather than calculation, on insight rather than formalism. The discussions are intended to be accessible to those without specialised training, while still pointing toward the deeper ideas that lie beneath the surface.

For those who wish to explore further, each chapter concludes with a bibliography, offering pathways into more rigorous treatments of the subject.

It is worth noting that mathematics, like any human endeavour, is shaped by the individuals who pursue it.

Throughout this book, we encounter mathematicians from different periods and cultures, including several from India whose contributions have been both profound and, at times, underappreciated. Their work reflects not only intellectual achievement but also a certain persistence—a willingness to engage with difficult ideas and follow them to their conclusions.

There is something quietly admirable about this.

Mathematics does not yield its insights easily. It requires patience, clarity of thought, and a tolerance for occasional confusion. Yet it rewards these efforts with a deeper understanding of the world and, perhaps, a sense of elegance that is difficult to find elsewhere.

This brings us to an aspect of mathematics that is often overlooked: its beauty.

To speak of beauty in mathematics may seem unusual, particularly to those who associate the subject with examinations rather than aesthetics. Yet mathematicians frequently describe their work in terms that would not be out of place in discussions of art or music.

An elegant proof, a simple equation that captures a complex idea, a pattern that reveals itself with unexpected clarity—these are sources of genuine delight.

The beauty of mathematics lies not only in its results but in its methods.

It transforms complexity into simplicity, uncertainty into structure, and disparate observations into coherent explanations. It does so with a precision that is both rigorous and, at times, surprisingly graceful.

There is also, one might say, a certain humour in mathematics.

Not the sort that provokes laughter in a crowded room, but a quieter form—a recognition of unexpected connections, of ideas that behave in ways that defy initial expectations. Infinity, for example, has a habit of producing results that are both perplexing and oddly amusing. Probability occasionally reveals that our intuitions are less reliable than we might hope.

These moments, while subtle, add a human dimension to the subject.

They remind us that mathematics is not merely a collection of facts, but an ongoing exploration—one that involves curiosity, creativity, and, occasionally, a sense of play.

If this book has a central aim, it is to invite the reader into that exploration.

It does not require prior expertise, only a willingness to consider ideas and follow them as they unfold. The journey moves from the familiar to the unfamiliar, from simple observations to deeper insights.

Along the way, one may begin to see mathematics not as an abstract discipline, but as a way of looking at the world.

A way of noticing patterns, of asking questions, and of finding connections.

And perhaps, by the end, the initial hesitation that often accompanies mathematics will have softened, replaced by a quiet appreciation.

For mathematics, when approached in this manner, is not something to be endured.

It is something to be enjoyed.

Preferably with a cup of tea, and a willingness to be occasionally surprised.

Dr Aloke Mullick

Pune, India

Prologue

Author's Note

It would be reassuring, for the sake of narrative symmetry, to say that I have always loved mathematics.

Unfortunately, that would be entirely untrue.

For a considerable portion of my academic life, mathematics and I maintained a relationship best described as distant and mutually unenthusiastic. I regarded it as a subject that demanded obedience rather than understanding, a sequence of procedures to be memorised and reproduced under examination conditions. It seemed less like a language of the universe and more like a particularly strict set of instructions issued without explanation.

Numbers, in those days, did not converse. They merely insisted.

Equations appeared not as expressions of relationships but as obstacles to be negotiated. Geometry was a collection of results to be accepted, algebra a tangle of symbols to be endured, and calculus—when it eventually arrived—seemed to possess an almost philosophical determination to be unapproachable.

I did what many reasonably sensible students do under such circumstances. I learned enough to proceed, avoided unnecessary engagement, and quietly concluded that mathematics was not for me.

It is a conclusion that many arrive at.

And it is, I have come to believe, a rather unfortunate one.

The turning point came, somewhat unexpectedly, late in college.

It was not a dramatic moment. There were no sudden revelations, no lightning bolts of insight, and certainly no immediate transformation into mathematical brilliance. Instead, there was a teacher.

A teacher who, with a combination of patience and clarity, did something quite remarkable: he treated mathematics not as a set of rules, but as a way of thinking.

He did not begin with formulas. He began with ideas.

Concepts that had previously seemed opaque were presented in a manner that made them almost obvious. Not simple, necessarily, but understandable. There was a sense that mathematics was not something imposed from the outside, but something that could be approached, explored, and, eventually, appreciated.

For the first time, the subject felt coherent.

More importantly, it felt human.

I began to see that behind every formula there was a reason, behind every method a logic, and behind every concept a story. Mathematics was not a collection of arbitrary procedures; it was a structure built with care, shaped by centuries of thought.

And, rather surprisingly, it possessed a certain elegance.

This was, for me, the beginning of a gradual change.

The resistance softened. The hesitation gave way to curiosity. Problems that once seemed intimidating became, if not easy, then at least approachable. There was even, on occasion, a sense of satisfaction—a rather unexpected development.

One might say that I began, quite cautiously at first, to enjoy mathematics.

This experience raises an interesting question.

How many people, I wonder, have formed their opinion of mathematics based not on the subject itself, but on how it was presented to them?

It is not a trivial matter.

Mathematics, perhaps more than many other disciplines, depends heavily on its mode of introduction. Presented as a series of rules to be memorised, it can appear rigid and uninviting. Presented as a system of ideas to be understood, it reveals a very different character.

The difference is not in the mathematics.

It is in the approach.

This book exists, in large part, because of that realisation.

It is an attempt to present mathematics not as an obligation, but as an invitation. Not as a test of endurance, but as an exploration. It does not seek to cover the subject exhaustively, nor to replace formal study. Instead, it aims to provide a perspective—a way of seeing mathematics that emphasises its connections, its elegance, and its quiet presence in the world.

The chapters that follow are written with this intention in mind.

They do not assume prior expertise. They do not dwell on technical detail for its own sake. Instead, they attempt to illuminate the ideas that lie beneath the surface, to show how mathematics connects with the sciences, with nature, and with everyday experience.

They also attempt, in a modest way, to capture the character of the subject.

Mathematics is often portrayed as austere, but it is, in fact, full of subtlety. It contains moments of surprise, instances of unexpected simplicity, and, occasionally, results that are almost amusing in their implications.

Infinity behaves in ways that defy intuition. Probability reveals that our instincts are not always reliable. Patterns appear where one might not expect them.

These are not merely technical details; they are part of what makes mathematics engaging.

Of course, it would be unrealistic to expect that every reader will develop a deep and abiding love for the subject.

Mathematics, like any field, has its own temperament, and not everyone will find it equally appealing. But if this book can achieve something more modest—if it can replace hesitation with curiosity, or uncertainty with a degree of appreciation—then it will have served its purpose.

And if, for some readers, it goes a little further—if it allows them to see mathematics not as a barrier but as a companion—then that would be particularly gratifying.

For mathematics, once understood in this way, is not something distant or inaccessible.

It is present in the patterns of nature, in the structure of scientific laws, in the rhythms of music, and in the logic of everyday reasoning. It is both a tool and a lens—a way of describing the world and a way of understanding it.

It does not demand perfection.

It asks only for attention.

And perhaps a little patience.

Looking back, it is difficult not to feel a certain gratitude toward that teacher who first made these ideas accessible.

The transformation was not immediate, nor was it complete. There are still areas of mathematics that I approach with caution, and moments when the subject retains its capacity to puzzle.

But the difference is this: what once seemed forbidding now appears intriguing.

What once felt like an obligation now feels like an opportunity.

That, I suspect, is the most one can reasonably hope for.

If this book can offer a similar shift in perspective—if it can open the door, even slightly, to the possibility that mathematics is not merely tolerable but, in its own quiet way, beautiful—then it will have justified its existence.

And if, somewhere along the way, a reader finds themselves unexpectedly enjoying a mathematical idea, or pausing to appreciate a pattern that might previously have gone unnoticed, then I shall consider the effort well spent.

After all, falling in love with mathematics need not be a dramatic affair.

It can begin, quite simply, with understanding.

And perhaps, with time, develop into something rather more enduring.

Dr Aloke Mullick

Pune, India

ONE

The Most Unreasonable Success of Mathematics

On a clear night, if one has the good fortune to be far away from city lights and mildly indifferent mosquitoes, it is possible to look upward and see an extraordinary spectacle. The sky, which during the day appears a rather bland shade of blue, suddenly reveals itself as a grand and rather crowded theatre. Stars appear everywhere—distant, ancient, and utterly unconcerned with the daily affairs of humanity.

What is remarkable is not merely that these celestial objects exist, but that their behaviour can be described with surprising accuracy using mathematics. With a handful of equations scribbled on paper, one can predict the motion of planets, the timing of eclipses, and even the path of distant comets that have not yet returned to visit us.

The universe, it seems, runs on mathematics with the quiet efficiency of a well-managed railway timetable.

This fact is both comforting and deeply peculiar.

Most human inventions behave exactly as expected. A toaster, for example, warms bread because someone designed it to do so. Mathematics, however, was not designed to run the universe. Mathematicians developed their ideas for reasons that were often abstract, playful, or occasionally obscure. Yet time and again, these very ideas turn out to describe nature with uncanny precision.

One might say that mathematics keeps turning up at the scene of scientific discovery like an enthusiastic amateur detective who somehow solves the mystery before the professionals arrive.

This curious phenomenon was noticed most famously by the physicist **Eugene Wigner**, who in 1960 wrote a celebrated essay titled *The Unreasonable Effectiveness of Mathematics in the Natural Sciences.* Wigner did not mean the phrase as a complaint, though the wording suggests a certain mild exasperation. His point was that the success of mathematics in explaining the physical world is astonishing—so astonishing, in fact, that it borders on the mysterious.

Why should equations invented in the quiet solitude of human thought end up describing the workings of stars, atoms, and galaxies?

No one is entirely certain.

To appreciate the strangeness of this situation, it is useful to remember that mathematics begins in rather humble circumstances. The earliest mathematics arose from everyday practical needs. Ancient farmers needed to measure fields. Merchants needed to count goods. Shepherds, perhaps the most patient accountants in history, needed to ensure that the number of sheep leaving the field in the morning matched the number returning in the evening.

These modest activities gave rise to numbers, and numbers gradually developed into arithmetic. Arithmetic, like a diligent clerk, performs its duties with quiet reliability. It adds, subtracts, multiplies, and divides with impeccable discipline.

Yet somewhere along the way, mathematics developed ambitions far beyond bookkeeping.

Consider geometry. The ancient Egyptians used geometric reasoning to measure land after the Nile floods erased property boundaries each year. The Greeks later transformed geometry into an intellectual masterpiece, particularly through the work of **Euclid**, whose *Elements* became one of the most influential books ever written.

Euclid's geometry was not concerned with land disputes or practical surveying. It was an elegant logical system built upon a few simple assumptions. Lines, triangles, and circles were studied not because they solved immediate problems but because they possessed a kind of intellectual beauty.

For centuries, geometry was admired as a purely abstract pursuit.

Then, quite unexpectedly, geometry returned to explain the universe itself.

In the early twentieth century, **Albert Einstein** discovered that gravity could be understood as a geometric property of space and time. According to his theory of general relativity, massive objects such as planets and stars bend the fabric of spacetime, and other objects move along the curves created by this bending.

In other words, gravity is geometry.

This revelation produced a curious situation in which ideas originally developed by ancient geometers turned out to describe the behaviour of galaxies. One suspects Euclid would have been both pleased and slightly bewildered.

The story repeats itself throughout the history of science.

During the seventeenth century, **Isaac Newton** developed calculus in order to describe motion and change. The mathematical tools he created allowed scientists to calculate the orbits of planets, the motion of projectiles, and the behaviour of fluids. These equations became the foundation of classical physics and remained astonishingly accurate for centuries.

What makes this particularly interesting is that calculus was initially developed through abstract reasoning about infinitesimal quantities. Newton did not discover calculus by observing falling apples and immediately writing down perfect equations. The mathematics emerged through intellectual exploration, guided partly by physical intuition but largely by the internal logic of mathematical thought.

Nature, it seems, had quietly arranged itself to follow these same mathematical rules.

The pattern continues into modern science. Quantum mechanics, which governs the behaviour of atoms and subatomic particles, relies on mathematical structures that are remarkably sophisticated. Much of the mathematics involved was developed long before physicists realised it would describe the physical world.

Indeed, physicists sometimes find themselves borrowing mathematical ideas that were invented decades earlier for entirely different reasons.

The situation occasionally resembles a library in which mathematicians have been quietly writing books for centuries, while scientists wander in later and discover that several of those books contain detailed instructions for running the universe.

This peculiar partnership between mathematics and nature raises an obvious question.

Is mathematics discovered, or is it invented?

If mathematics is invented, then its success in describing the universe seems almost miraculous. It would be as if someone invented the rules of chess and later discovered that the planets had been following those rules all along.

On the other hand, if mathematics is discovered—if it represents a deeper structure underlying reality—then its effectiveness becomes less surprising. In that case, mathematicians are not inventing mathematics but uncovering patterns that already exist in the fabric of the universe.

The debate has occupied philosophers and mathematicians for centuries, and it shows no sign of being resolved soon.

Fortunately, the universe appears to function quite happily regardless of our philosophical uncertainties.

One of the most delightful aspects of mathematics is its ability to describe patterns in nature that might otherwise go unnoticed. The spirals of seashells, the branching of trees, the arrangement of leaves around a stem—these structures often follow mathematical rules with remarkable consistency.

Even something as simple as a sunflower reveals hidden mathematics. If one examines the arrangement of seeds within the flower, one finds spiral patterns that correspond to a famous sequence of numbers known as the Fibonacci sequence. The seeds organise themselves in a way that maximises efficiency, packing together with elegant precision.

Nature, it appears, is fond of mathematical optimisation.

Of course, mathematics is not always obvious at first glance. Many natural phenomena appear chaotic or unpredictable. Weather systems, for example, have a reputation for ignoring even the most confident forecasts.

Yet even chaos turns out to possess mathematical structure.

In the late twentieth century, mathematicians discovered that chaotic systems often follow deterministic rules that produce complex patterns. This field, known as chaos theory, revealed that seemingly random behaviour can emerge from simple mathematical equations.

The discovery was both humbling and reassuring. Humbling, because it showed that simple rules can produce astonishing complexity. Reassuring, because even the most unruly systems still obey mathematics.

The deeper scientists look into the workings of the universe, the more mathematics they find waiting there.

Atoms obey mathematical equations. Light follows mathematical laws. The expansion of the universe itself appears to be governed by elegant mathematical relationships.

One begins to suspect that mathematics is not merely a convenient tool for describing reality, but something woven into its very structure.

Why this should be so remains one of the great intellectual mysteries.

Some thinkers have suggested that the universe is fundamentally mathematical in nature. Others propose that the human mind, having evolved within the universe, naturally develops mathematical concepts that mirror its underlying patterns.

Both possibilities are intriguing. Both are also slightly unsettling.

After all, if the universe truly runs on mathematics, then somewhere among the equations and symbols lies the blueprint of reality itself.

Fortunately for the rest of us, mathematicians seem content to pursue these ideas with quiet enthusiasm rather than alarming ambition.

They sit in offices and libraries, scribbling symbols, proving theorems, and occasionally arguing about definitions. From the outside, their work may appear detached from everyday life.

Yet history repeatedly shows that these abstract explorations eventually find their way into the practical world.

Number theory, once considered the most useless branch of mathematics, now forms the foundation of modern cryptography. Geometry describes the structure of spacetime. Probability governs fields as diverse as medicine, economics, and climate science.

Mathematics, in short, keeps proving itself useful in places where no one initially expected it to be.

This brings us back to Wigner's observation about the unreasonable effectiveness of mathematics.

Perhaps the most remarkable aspect of mathematics is not merely that it works, but that it works so elegantly. Simple equations often describe complex phenomena with astonishing precision. Entire branches of science rest upon mathematical relationships that can be written on a single line.

There is a certain quiet beauty in this simplicity.

The universe could have been messy and incoherent, refusing to cooperate with human attempts at understanding. Instead, it reveals patterns that mathematics captures with surprising grace.

Why this should be the case remains, for now, an open question.

But it is a rather pleasant mystery to contemplate—preferably on a quiet evening, under a sky filled with stars that continue to follow their mathematical paths with admirable discipline.

One suspects they have been doing so for quite some time.

Bibliography – Chapter 1

Wigner, Eugene. *The Unreasonable Effectiveness of Mathematics in the Natural Sciences.* Communications in Pure and Applied Mathematics, 1960.

Penrose, Roger. *The Road to Reality: A Complete Guide to the Laws of the Universe.* Jonathan Cape, 2004.

Stewart, Ian. *Nature's Numbers: The Unreal Reality of Mathematical Imagination.* Basic Books, 1995.

Hardy, G. H. *A Mathematician's Apology.* Cambridge University Press, 1940.

Tao, Terence. *Solving Mathematical Problems: A Personal Perspective.* Oxford University Press, 2006.

TWO

COUNTING, CURIOSITY AND THE DAWN OF NUMBERS

It is generally agreed that civilisation began when someone, somewhere, decided it would be prudent to keep track of things.

Quite what those things were is open to interpretation. One suspects they were not initially philosophical concepts or astronomical observations, but rather items of a more pressing nature—grain, livestock, tools, and perhaps the occasional neighbour's goat whose ownership was, under certain circumstances, a matter of lively debate.

Counting, then, began not as an intellectual pursuit but as a practical necessity. It is difficult to run even the simplest society without some notion of quantity. A farmer who cannot distinguish between three sacks of grain and thirty is unlikely to enjoy a long or prosperous career.

And so, long before mathematics acquired its present air of abstraction and quiet dignity, it existed in a rather earthy form: tally marks scratched into bone, notches carved into wood, pebbles arranged in small, reassuring piles.

One of the earliest known artefacts of this numerical instinct is the Ishango bone, discovered in central Africa, which bears a series of notches that appear to represent some form of counting. It is not, perhaps, the most elegant mathematical text ever produced, but it demonstrates a crucial point: the human mind, even in its earliest stages of cultural development, possessed an irrepressible urge to quantify.

This urge, as it turns out, would change the course of history.

At first, numbers were inseparable from the objects they described. Three sheep were not an abstract concept; they were quite literally three sheep, preferably within sight. The idea that "three" could exist independently of sheep, stones, or spoons required a subtle leap of abstraction—one that humanity took gradually and, one imagines, with a certain amount of hesitation.

It is one thing to count apples. It is quite another to contemplate the number "three" as an entity in its own right.

Yet this step, modest as it may seem, marks the true beginning of mathematics.

Different civilisations approached this problem with varying degrees of ingenuity. The ancient Egyptians, for example, developed a system of numerals that allowed them to represent numbers using hieroglyphic symbols. It was perfectly serviceable, though not particularly efficient by modern standards. Writing large numbers required a certain amount of patience and, ideally, a steady hand.

The Mesopotamians, on the other hand, adopted a rather more ambitious approach. They developed a base-60 number system, remnants of which survive today in the way we measure time and angles. Sixty seconds in a minute, sixty minutes in an hour—these are quiet echoes of a civilisation that flourished several thousand years ago.

Why base 60? The exact reason remains somewhat unclear, though it has the advantage of being divisible by many numbers, which is a quality mathematicians tend to admire.

Meanwhile, in other parts of the world, counting systems evolved along different lines. The Chinese developed rod numerals and sophisticated methods for calculation. The Mayans devised a vigesimal (base-20) system, complete with a symbol for zero, which we shall return to shortly, as it deserves special attention.

And then, in the Indian subcontinent, something rather extraordinary occurred.

Numbers, which had thus far been somewhat cumbersome and tied to specific representations, underwent a quiet but profound transformation. Indian mathematicians developed a positional decimal system—a method of representing numbers using ten symbols, where the position of each symbol determines its value.

This may sound unremarkable to the modern reader, who has been using such a system since childhood, often without much enthusiasm. Yet it is one of the most important intellectual developments in human history.

The brilliance of the system lies in its simplicity. With just ten digits, one can represent any number, no matter how large. More importantly, arithmetic operations become vastly more efficient.

One can, for instance, add or multiply numbers without resorting to elaborate symbols or unwieldy notations. The system is, in a word, elegant.

At the heart of this innovation lies a concept so deceptively simple that it is often taken for granted: zero.

The idea of zero—the representation of nothingness as a number—is one of humanity's most remarkable achievements. It is also, if one reflects upon it for too long, slightly unsettling.

After all, how does one represent nothing?

It was in India that zero first emerged not merely as a placeholder but as a number in its own right, with defined mathematical properties. The mathematician **Brahmagupta**, writing in the seventh century, provided rules for arithmetic involving zero, treating it with the same seriousness as any other number.

This was no small matter.

Zero allows the positional number system to function. Without it, distinguishing between, say, 105 and 15 becomes rather awkward. With it, numbers gain clarity, structure, and an almost architectural elegance.

One might say that zero is the quiet hero of mathematics—rarely celebrated, but absolutely indispensable.

From India, this numerical system travelled westward, carried by scholars and traders. It was eventually transmitted to the Islamic world, where mathematicians refined and expanded upon it. The Persian scholar **Al-Khwarizmi** played a crucial role in this process, writing works that introduced these numerical methods to a broader audience.

His name, incidentally, gives us the word "algorithm," which is a pleasing reminder that even the most modern computational concepts have rather ancient origins.

The system later reached Europe, where it was initially met with a mixture of curiosity and suspicion. Roman numerals, though undeniably dignified, were not particularly well suited to complex calculations. The new system, with its unfamiliar symbols and unsettling reliance on zero, seemed at first somewhat exotic.

Gradually, however, practicality prevailed.

Merchants found the new numerals far more convenient for bookkeeping. Mathematicians appreciated their efficiency. Over time, the so-called "Arabic numerals"—which were, in fact, of Indian origin—became the standard numerical system across much of the world.

And thus, almost without fanfare, the modern number system took its place at the foundation of science, commerce, and daily life.

It is worth pausing to consider how extraordinary this development is.

With ten simple symbols—0 through 9—we can describe quantities ranging from the number of apples in a basket to the number of atoms in the observable universe. We can measure distances between galaxies, calculate the energy of stars, and keep track of one's monthly expenses, all using the same basic framework.

Few human inventions possess such versatility.

Yet numbers are not merely tools for counting. They also reveal patterns—patterns that seem to exist independently of human intention.

Consider the sequence of natural numbers: 1, 2, 3, 4, and so on. At first glance, it appears straightforward enough. Yet within this sequence lie structures that have fascinated mathematicians for centuries.

Prime numbers, for instance, are numbers that can only be divided by 1 and themselves. They appear irregularly among the natural numbers, like solitary figures in a crowd. Despite their apparent randomness, they obey subtle patterns that mathematicians continue to explore.

There is something almost poetic about prime numbers. They are simple to define, yet their distribution remains one of the deepest mysteries in mathematics.

Then there are sequences such as the Fibonacci numbers, where each term is the sum of the two preceding ones. These numbers appear in nature with surprising frequency, governing patterns of growth in plants and the arrangement of leaves and seeds.

It is as though numbers, once discovered, begin to reveal a hidden order underlying the world.

This brings us back to the role of curiosity.

Mathematics, despite its practical origins, has always been driven by a certain intellectual restlessness. People do not merely count things because they must; they also count because they are curious.

What happens if numbers continue indefinitely? Are there infinitely many primes? Can numbers describe motion, shape, or chance?

These questions, initially abstract, often lead to profound discoveries.

The history of mathematics is filled with individuals who pursued such questions not for immediate practical benefit, but for the sheer pleasure of understanding. Their work, in turn, laid the foundations for later applications.

One is reminded of the peculiar habit mathematicians have of solving problems that no one has yet realised are important.

And yet, time and again, those very problems turn out to be essential.

In modern life, numbers are everywhere. They appear on clocks, in bank accounts, on mobile phones, and in the invisible calculations performed by computers. We rely on them so completely that we rarely pause to consider their origin.

It is rather like breathing—essential, continuous, and generally unnoticed.

But behind these familiar symbols lies a long and fascinating history, stretching back thousands of years. It is a story of gradual abstraction, of increasing sophistication, and of a growing realisation that numbers are not merely practical tools but windows into deeper patterns.

From tally marks scratched into bone to the elegant decimal system, from the hesitant recognition of "three" as an abstract concept to the confident manipulation of enormous quantities, the journey of numbers reflects the evolution of human thought itself.

There is, perhaps, something quietly profound in this.

Numbers begin with the simplest of observations—there is one object, then another, then another. Yet from this humble beginning arises an entire universe of ideas.

It is tempting to think of mathematics as something distant and difficult, accessible only to specialists armed with advanced training and considerable patience. But at its heart, mathematics begins with something much simpler.

It begins with noticing.

Noticing that quantities can be compared. Noticing that patterns repeat. Noticing that relationships exist between things that might otherwise seem unrelated.

Counting, in this sense, is not merely a technical skill. It is an expression of curiosity—a way of making sense of the world.

And once that curiosity is awakened, it tends to lead further than one might expect.

From counting sheep to contemplating infinity is, admittedly, a considerable journey. But it is one that humanity has undertaken with remarkable success.

Along the way, numbers have evolved from practical tools into objects of study in their own right. They have acquired properties, relationships, and behaviours that continue to intrigue and occasionally confound those who study them.

And yet, despite all this sophistication, numbers retain their original simplicity.

One can still count on one's fingers.

This, perhaps, is the most charming aspect of mathematics. It begins with the most ordinary of experiences and gradually unfolds into something vast and intricate.

The child learning to count and the mathematician exploring abstract number theory are, in a sense, engaged in the same activity. Both are seeking to understand patterns, to impose order, and to satisfy a certain intellectual curiosity.

It is a rather reassuring continuity.

As we move forward, numbers will become more sophisticated, their relationships more intricate, and their applications more surprising. But it is worth remembering that all of this rests upon a simple foundation.

Someone, long ago, looked at a collection of objects and thought, "It might be useful to keep track of these."

From that modest beginning arose one of the greatest intellectual achievements in human history.

Not bad, one might say, for a few scratches on a bone.

Bibliography – Chapter 2

Ifrah, Georges. *The Universal History of Numbers.* Wiley, 2000.

Plofker, Kim. *Mathematics in India.* Princeton University Press, 2009.

Joseph, George Gheverghese. *The Crest of the Peacock: Non-European Roots of Mathematics.* Princeton University Press, 2011.

Datta, B. and Singh, A. N. *History of Hindu Mathematics.* Asia Publishing House, 1962.

Brahmagupta. *Brahmasphutasiddhanta* (translated editions).

Katz, Victor J. *A History of Mathematics: An Introduction.* Addison-Wesley, 2008.

Menninger, Karl. *Number Words and Number Symbols.* MIT Press, 1992.

THREE

GEOMETRY: THE UNIVERSE DRAWN WITH A STRAIGHTEDGE

There are, broadly speaking, two ways of looking at the world.

One may observe it as it appears—messy, irregular, and occasionally inconvenient—or one may suspect, quietly and with a hint of optimism, that beneath this apparent disorder lies a certain structure. Geometry belongs firmly to the latter approach. It is the art of finding order in space, of drawing neat lines where nature has provided only approximation, and of insisting—politely but firmly—that shapes ought to behave themselves.

The origins of geometry are, like those of most respectable disciplines, somewhat humble. One does not begin with triangles and theorems, but with land. Specifically, with land that has inconveniently shifted, flooded, or otherwise refused to remain where it was last measured.

In ancient Egypt, the annual flooding of the Nile was both a blessing and a mild administrative nuisance. While the floodwaters enriched the soil, they also erased boundary markers, leaving farmers in the delicate position of having to renegotiate whose field was whose. It was in this context that geometry—literally "earth measurement"—first took on practical importance.

Surveyors, armed with ropes and a commendable sense of patience, began to measure distances and reconstruct boundaries. They discovered, perhaps with some satisfaction, that certain shapes behaved predictably. A triangle, for example, was far less argumentative than a group of neighbouring farmers.

From these practical beginnings, geometry slowly evolved into something more refined.

If the Egyptians used geometry as a tool, the Greeks transformed it into an art form. They approached geometry not merely as a means of measurement but as a system of reasoning. It was no longer sufficient to know that a particular construction worked; one had to understand why it worked, and ideally, to prove it with a certain degree of elegance.

This intellectual shift reached its most celebrated expression in the work of **Euclid**, whose *Elements* remains one of the most influential books ever written. Euclid did something rather remarkable. He took a collection of geometric knowledge and organised it into a logical system built upon a small set of assumptions, or axioms.

From these simple beginnings, he derived an entire universe of geometric truths.

The method is deceptively straightforward. One begins with a few basic statements—such as the idea that a straight line can be drawn between any two points—and then proceeds, step by step, to prove increasingly complex results. Each theorem follows logically from the previous ones, creating a structure that is both rigorous and, in its own way, rather beautiful.

It is worth noting that Euclid's geometry is not concerned with the imperfections of the real world. The lines he describes have no thickness. The points have no size. Circles are perfectly round, and triangles behave with impeccable discipline.

Reality, of course, is somewhat less cooperative. Lines drawn on paper wobble slightly. Circles are never perfectly symmetrical. Triangles, when constructed by hand, occasionally display a mild reluctance to conform to theoretical expectations.

Yet this is precisely the strength of geometry. By abstracting away the imperfections of the physical world, it reveals underlying principles that remain true regardless of circumstance.

A triangle, in Euclidean geometry, has interior angles that sum to 180 degrees. This is not a matter of approximation or convenience; it is a logical necessity, derived from the axioms of the system.

There is something deeply satisfying about this certainty.

Geometry, in this sense, offers a glimpse of a world governed not by chance or approximation, but by reason. It is a place where statements can be proven with absolute clarity, where ambiguity is gently but firmly excluded.

One begins to see why the Greeks regarded geometry as a pathway to philosophical insight.

Yet geometry did not develop solely in Greece. In the Indian subcontinent, geometric ideas emerged in a rather different context. The **Sulba Sutras**, ancient texts concerned with the construction of ritual altars, contain sophisticated geometric rules. These include methods for constructing right angles and transformations between shapes of equal area.

The purpose was not abstract reasoning but precise construction. Altars had to be built according to strict specifications, and geometry provided the necessary tools.

It is a pleasing example of how practical necessity and intellectual curiosity often walk hand in hand.

Among the geometric insights found in these texts is an early understanding of what is now known as the Pythagorean theorem—the relationship between the sides of a right-angled triangle. While the theorem is most commonly associated with **Pythagoras**, similar ideas appeared independently in multiple cultures.

This, one suspects, is the sort of thing that happens when people begin drawing triangles with sufficient enthusiasm.

As geometry developed, it gradually expanded beyond the study of simple shapes. Circles, polygons, and conic sections—ellipses, parabolas, and hyperbolas—entered the picture. These curves, initially studied for their own sake, would later find applications in fields as diverse as astronomy and engineering.

The ellipse, for instance, turned out to describe the orbits of planets, as discovered by Johannes Kepler. One imagines the ellipse felt rather pleased with itself upon learning this.

At this point, geometry begins to display a characteristic that will become increasingly familiar: ideas developed in abstraction often return to explain the physical world.

For many centuries, Euclidean geometry was regarded as the definitive description of space. It seemed self-evident that parallel lines never meet, that the angles of a triangle sum to 180 degrees, and that space behaves in a reassuringly straightforward manner.

And then, in the nineteenth century, mathematicians began to ask an awkward question.

What if this is not always the case?

By modifying Euclid's fifth postulate—the statement concerning parallel lines—mathematicians developed entirely new geometries in which the familiar rules no longer applied. In these non-Euclidean geometries, triangles could have angle sums greater than or less than 180 degrees, and parallel lines behaved in ways that would have made Euclid raise an eyebrow.

At first, these geometries were regarded as curious intellectual exercises. They were interesting, certainly, but not particularly relevant to the real world.

Then along came **Albert Einstein**, who had a habit of taking abstract ideas and applying them to reality with disconcerting effectiveness.

In his theory of general relativity, Einstein proposed that space itself is curved, and that this curvature is what we perceive as gravity. Massive objects distort the geometry of spacetime, causing other objects to move along curved paths.

In this framework, non-Euclidean geometry is not merely a mathematical curiosity; it is the language of the universe.

Triangles drawn on the surface of a sphere, for example, do not obey the familiar rules of Euclidean geometry. Their angles can sum to more than 180 degrees. This is precisely the kind of geometry required to describe the curved spacetime around massive objects.

Once again, mathematics—developed in abstraction—finds itself at the heart of physical reality.

It is difficult not to admire the quiet audacity of this arrangement.

Geometry, which began with ropes and fields, now describes the structure of the cosmos.

Of course, geometry is not confined to the grand scale of galaxies and spacetime. It is equally present in the everyday world, though often in less dramatic fashion.

Consider architecture. Buildings, bridges, and roads all rely on geometric principles. The stability of a structure depends on the arrangement of its components, the distribution of forces, and the relationships between angles and lengths.

A well-designed bridge is, in essence, a geometric solution to a practical problem.

Art, too, makes frequent use of geometry. Perspective, symmetry, and proportion all rely on geometric ideas. The Renaissance artists, in particular, developed sophisticated techniques for representing three-dimensional space on a two-dimensional surface.

The result is a curious collaboration between mathematics and aesthetics. Geometry provides the structure; art provides the expression.

Even nature, which is often accused of being untidy, exhibits a surprising fondness for geometric patterns. Snowflakes display intricate symmetry. Honeycombs form hexagonal grids of remarkable efficiency. Crystals grow in regular, repeating structures.

One might almost suspect that nature has been quietly studying geometry in its spare time.

Yet geometry is not merely about recognising patterns; it is also about creating them.

With a straightedge and compass—tools of admirable simplicity—one can construct an astonishing variety of shapes. Bisecting angles, drawing perpendicular lines, inscribing circles—these operations form the basis of classical geometric construction.

There is something almost meditative about these activities. The act of drawing a precise line or constructing a perfect circle carries a certain satisfaction, as though one is participating in a small but meaningful act of order.

It is perhaps no coincidence that geometry has long been associated with education. Learning geometry is not merely about acquiring knowledge; it is about developing a way of thinking. It teaches precision, logical reasoning, and an appreciation for structure.

At the same time, geometry retains a sense of playfulness.

Problems involving shapes and constructions often resemble puzzles, inviting exploration and creativity. There is a quiet delight in discovering that a seemingly complex figure can be understood through a simple idea.

This balance between rigour and play is one of geometry's most appealing qualities.

As mathematics has evolved, geometry has expanded into new and unexpected areas. Topology, for example, studies properties of shapes that remain unchanged under continuous deformation. In this curious branch of mathematics, a coffee cup and a doughnut are considered equivalent, as both have a single hole.

This may seem slightly absurd at first, but it reflects a deeper principle: geometry is not limited to rigid shapes. It encompasses a wide range of structures, from the familiar to the abstract.

And yet, despite all this sophistication, the essence of geometry remains remarkably simple.

It is the study of space and shape.

It begins with points and lines, with the act of drawing and measuring. From these humble beginnings, it builds a framework that extends from the smallest constructions to the vast structure of the universe.

There is a certain elegance in this progression.

One starts with a straightedge and ends with spacetime.

It is, perhaps, one of the most compelling examples of how mathematics transforms simple ideas into profound insights. Geometry does not overwhelm with complexity; it reveals, step by step, the underlying order of the world.

And in doing so, it offers a quiet reassurance.

The universe, for all its apparent chaos, is not entirely without structure. Beneath the irregularities and uncertainties lies a framework that can be understood, described, and, with a little effort, appreciated.

All it requires is a line, a point, and a willingness to follow them wherever they may lead.

Bibliography – Chapter 3

Euclid. *Elements.* Translated editions, various publishers.

Heath, T. L. *The Thirteen Books of Euclid's Elements.* Cambridge University Press.

Datta, B. and Singh, A. N. *History of Hindu Mathematics.* Asia Publishing House.

Sarasvati Amma, T. A. *Geometry in Ancient and Medieval India.* Motilal Banarsidass.

Greenberg, Marvin Jay. *Euclidean and Non-Euclidean Geometries.* W. H. Freeman.

Penrose, Roger. *The Road to Reality.* Jonathan Cape.

Hilbert, David. *Foundations of Geometry.* Open Court Publishing.

FOUR

Algebra: When Numbers Learn to Converse

If arithmetic is the polite business of dealing with numbers as they are, algebra is what happens when numbers develop ambitions.

In arithmetic, numbers behave rather like well-trained civil servants. They add, subtract, multiply, and divide without complaint. Their identities are fixed, their duties clear, and their conduct, on the whole, exemplary.

Algebra, however, introduces a subtle but transformative idea: numbers need not always reveal themselves immediately. They may, if they wish, conceal their identity behind symbols—letters, no less—and still participate fully in mathematical discourse.

This is rather like attending a dinner party where half the guests decline to disclose their names, yet insist on being included in the conversation.

At first glance, this may seem unnecessarily complicated. After all, why replace a perfectly good number with an enigmatic symbol such as x? One suspects that many a student, upon first encountering algebra, has wondered whether this is not an elaborate joke played by mathematicians upon the unsuspecting.

Yet the idea, once understood, proves to be remarkably powerful.

Arithmetic answers specific questions. Algebra, by contrast, answers entire classes of questions at once. It does not merely tell you that two apples plus three apples make five apples; it tells you that *any* two quantities, when combined in a particular way, will behave according to a general rule.

In this sense, algebra is less concerned with individual facts and more with relationships.

It is, one might say, the grammar of mathematics.

To appreciate how this came about, it is useful to return briefly to the practical origins of mathematics. Early civilisations, having mastered the art of counting, soon encountered more complicated problems. Trade, taxation, land measurement, and astronomy all required calculations that could not always be resolved by simple arithmetic.

Suppose, for example, that one knows the total value of a collection of goods and the price of individual items, but not the number of items themselves. This is not an uncommon situation, particularly in markets where bargaining is conducted with enthusiasm.

To solve such problems, one must reason about unknown quantities.

Initially, this reasoning was expressed in words. Ancient texts from Egypt and Mesopotamia contain problems phrased in a narrative style, describing quantities and their relationships in sentences that would not be entirely out of place in a modestly perplexing word problem.

Over time, however, mathematicians began to seek more efficient ways of representing these relationships.

In the ancient world, one of the earliest systematic approaches to algebra appeared in the work of the Persian mathematician **Al-Khwarizmi**. His treatise, written in the ninth century, laid out methods for solving equations in a clear and methodical manner.

The title of his work, *Al-Jabr*, gives us the word "algebra," which is a pleasing reminder that the subject has a history far richer than its occasional reputation for causing distress in classrooms.

Al-Khwarizmi's approach was rhetorical—that is, it described equations in words rather than symbols. A typical problem might involve dividing a quantity into parts, recombining them, and determining an unknown value through logical steps.

While this method was effective, it was not particularly concise. One could solve problems, certainly, but one had to wade through a fair amount of prose in the process.

The true transformation of algebra occurred when mathematicians began to adopt symbolic notation.

Instead of describing unknown quantities in words, they represented them with letters. Instead of explaining operations at length, they used symbols such as +, –, and =.

This shift may appear modest, but it dramatically changed the nature of mathematics.

With symbolic notation, equations became compact, precise, and, perhaps most importantly, manipulable. One could rearrange them, combine them, and generalise them with ease.

Numbers, having been granted symbolic representation, could now engage in a form of conversation.

Consider a simple equation:

$x + 3 = 7$

Here, x represents an unknown quantity. The equation asserts a relationship: whatever x may be, when three is added to it, the result is seven.

The task is to determine the identity of x.

In solving this equation, one performs a sequence of logical steps. Subtract three from both sides, and one finds that $x = 4$.

This process may seem straightforward, but it embodies a powerful principle: equality is preserved under certain operations. One can manipulate both sides of an equation in the same way without altering its truth.

This idea allows algebra to function as a system of reasoning.

More importantly, it allows one to solve problems without knowing the answer in advance.

In arithmetic, one typically begins with known quantities and computes a result. In algebra, one begins with a relationship and deduces the unknown.

It is, in a sense, a reversal of perspective.

As algebra developed, mathematicians began to explore increasingly complex equations. Quadratic equations, involving squares of unknowns, became a central focus. Methods for solving these equations were refined across cultures, including in India, where mathematicians such as **Bhaskara II** made significant contributions.

Bhaskara's work included elegant methods for solving equations and a keen appreciation for mathematical structure. His writings often combined technical insight with a certain poetic flair, suggesting that algebra, even in its early stages, possessed an aesthetic dimension.

This is perhaps not surprising.

Algebra, at its best, exhibits a kind of symmetry and balance. Equations can be rearranged in multiple ways, yet they retain their underlying truth. Solutions emerge not through brute force but through careful reasoning.

There is, one might say, a certain conversational elegance to the process.

As the subject progressed, algebra began to move beyond solving specific types of equations and toward a more general study of structures.

Polynomials, functions, and systems of equations entered the scene. Mathematicians developed techniques for handling expressions of increasing complexity, often discovering patterns that held across entire classes of problems.

At this point, algebra begins to resemble a language in its own right.

Just as language allows us to express ideas about the world, algebra allows us to express relationships between quantities. It provides a framework for describing change, interaction, and dependency.

This makes it indispensable to science.

In physics, for example, algebra is used to describe motion, forces, and energy. Equations relate variables such as velocity, acceleration, and time, allowing scientists to predict the behaviour of physical systems.

In economics, algebra models supply and demand, costs and revenues. In biology, it describes population growth and genetic inheritance.

One begins to see that algebra is not merely a branch of mathematics; it is a universal tool for understanding relationships.

It is, if one may put it this way, the quiet negotiator of the scientific world.

Yet algebra is not without its quirks.

For one thing, it has a tendency to become increasingly abstract. As mathematicians explored its possibilities, they began to consider objects that were not numbers in the traditional sense. Variables could represent functions, matrices, or even more exotic entities.

This led to the development of modern algebra, which studies structures such as groups, rings, and fields. These concepts, while undeniably important, have a habit of sounding slightly mysterious to the uninitiated.

One might encounter statements such as “this set forms a group under the given operation,” which, while perfectly clear to a mathematician, may strike others as the beginning of a rather exclusive club.

Nevertheless, these abstractions have proved immensely useful.

Group theory, for example, plays a crucial role in physics, particularly in the study of symmetry. It helps explain why certain physical laws remain unchanged under specific transformations.

In chemistry, algebraic structures describe molecular symmetries. In computer science, they underpin algorithms and cryptographic systems.

Once again, algebra demonstrates its remarkable ability to extend far beyond its original scope.

Despite this sophistication, the essence of algebra remains accessible.

At its heart, algebra is about recognising patterns and expressing relationships. It is about understanding how one quantity depends on another, and how changes in one affect the other.

This is not, in itself, an esoteric pursuit.

Indeed, one might argue that algebraic thinking is a natural extension of everyday reasoning. When one considers how expenses relate to income, or how travel time depends on speed and distance, one is engaging in a form of algebra.

The difference lies primarily in the level of abstraction.

Algebra takes these intuitive ideas and expresses them in a precise, systematic way. It replaces vague notions with clear relationships, allowing for rigorous analysis and prediction.

It is, in short, a way of thinking.

There is also a certain quiet pleasure in mastering algebra.

An equation that initially appears complicated can, with a few well-chosen steps, be transformed into something simple and comprehensible. The process is rather like untangling a knot: one begins with confusion and ends with clarity.

This transformation—from complexity to simplicity—is one of the defining features of mathematics.

It is also one of its most satisfying aspects.

Of course, algebra has acquired a somewhat mixed reputation over the years. For many students, it represents the moment when mathematics ceases to be straightforward and begins to ask awkward questions.

Why must numbers be replaced with letters? Why do equations need to be rearranged? And what, precisely, is the point of it all?

These are reasonable questions, and they deserve reasonable answers.

The point of algebra is not merely to solve equations, though it does this admirably. Its true purpose is to provide a framework for understanding relationships.

Without algebra, much of modern science would be impossible. We would lack the tools to describe motion, predict outcomes, and model complex systems.

In a very real sense, algebra allows us to make sense of the world.

It teaches us that beneath apparent complexity lies structure, and that this structure can be understood through careful reasoning.

It also teaches a certain intellectual discipline.

Algebra requires one to follow logical steps, to maintain consistency, and to respect the rules of the system. It rewards precision and punishes carelessness, though usually in a manner that is more instructive than vindictive.

One might even say that algebra encourages good manners in thought.

As we continue our journey through mathematics, algebra will remain a constant companion. It will appear in various forms, sometimes familiar, sometimes slightly disguised, but always performing the same essential function.

It will connect ideas, reveal patterns, and provide a language for expressing relationships.

And, on occasion, it will remind us that numbers, when given the opportunity, are perfectly capable of holding a conversation.

One need only listen carefully.

Bibliography – Chapter 4

Katz, Victor J. *A History of Mathematics: An Introduction*. Addison-Wesley, 2008.

Boyer, Carl B. *A History of Mathematics*. Wiley, 1991.

Al-Khwarizmi. *The Compendious Book on Calculation by Completion and Balancing (Al-Jabr)*. Translated editions.

Joseph, George Gheverghese. *The Crest of the Peacock*. Princeton University Press, 2011.

Datta, B. and Singh, A. N. *History of Hindu Mathematics*. Asia Publishing House, 1962.

Bhaskara II. *Lilavati* and *Bijaganita* (translated editions).

Artin, Emil. *Algebra*. Prentice Hall.

FIVE

Infinity, the Number That Refuses to Behave

There are certain ideas in mathematics that behave impeccably. The number two, for instance, is a model citizen. It is even, predictable, and rarely given to philosophical outbursts. Three is similarly well-mannered, though perhaps a touch more interesting. Even fractions, despite their occasional untidiness, generally conduct themselves within acceptable limits.

Infinity, however, is another matter entirely.

Infinity is the sort of concept that enters the room uninvited, rearranges the furniture, and then insists that the house has always been arranged that way. It resists ordinary intuition, ignores conventional rules, and has a disconcerting tendency to produce results that seem, at first glance, quite impossible.

And yet, mathematics would be considerably poorer without it.

To understand infinity, one must first appreciate a simple observation: numbers do not appear to end. One can count indefinitely—1, 2, 3, 4, and so on—without ever reaching a final number. This seemingly trivial fact contains the seed of a profound idea.

There is no largest number.

This realisation, which has likely occurred to countless individuals at various points in history (often during moments of mild distraction), leads naturally to the concept of infinity. If numbers continue without bound, then there is, in some sense, an "infinite" quantity.

Of course, stating that something goes on forever is one thing; understanding what that means is quite another.

Early thinkers approached infinity with a mixture of fascination and caution. In ancient Greece, philosophers such as Zeno of Elea devised paradoxes that highlighted the peculiar nature of infinite processes. Zeno's most famous example involves a race between the hero Achilles and a tortoise.

In this race, Achilles gives the tortoise a head start. By the time Achilles reaches the point where the tortoise began, the tortoise has moved slightly ahead. When Achilles reaches this new point, the tortoise has again advanced a little further.

This process, Zeno argued, continues indefinitely. Achilles, it would seem, can never quite catch the tortoise.

The conclusion is, of course, absurd. In reality, Achilles would overtake the tortoise with little difficulty, assuming he is in reasonably good form. Yet the paradox reveals something important: infinite sequences can produce results that defy immediate intuition.

Resolving such paradoxes required a more careful understanding of infinity—one that would take many centuries to develop.

Meanwhile, in the Indian philosophical tradition, infinity was approached with a rather different temperament. Instead of treating it as a source of paradox, it was often regarded as a natural and even essential aspect of reality.

The Upanishadic idea that the infinite remains unchanged even when something is taken from it—"that is whole; this is whole"—suggests an intuitive grasp of a property that would later appear in mathematics: infinity does not behave like ordinary numbers.

This is, one might say, a remarkably calm response to a rather unruly concept.

In mathematical terms, infinity begins to take on a more precise meaning with the development of calculus. In the seventeenth century, **Isaac Newton** and **Gottfried Wilhelm Leibniz** independently developed methods for dealing with quantities that approach zero or grow without bound.

Calculus does not treat infinity as a number in the usual sense. Instead, it considers limits—how quantities behave as they become arbitrarily large or small. This allows mathematicians to work with infinite processes in a controlled and rigorous way.

For example, one can consider the sum of an infinite series:

$1 + 1/2 + 1/4 + 1/8 + \ldots$

At first glance, this sum appears to grow without bound. After all, there are infinitely many terms. Yet, somewhat surprisingly, the sum converges to a finite value—namely, 2.

This result is both elegant and faintly unsettling.

It suggests that an infinite process can produce a finite outcome, which is not the sort of behaviour one expects from everyday experience. If one were to eat half a cake, then half of what remains, and so on indefinitely, one might reasonably hope to consume the entire cake. Mathematics confirms that this is indeed possible, though it does not comment on the practicalities.

As calculus developed, mathematicians became increasingly comfortable with infinity, though they continued to treat it with a certain degree of respect. Infinity was useful, certainly, but it required careful handling.

Then, in the nineteenth century, a mathematician arrived who decided that infinity deserved a more thorough investigation.

This was **Georg Cantor**, whose work on set theory transformed the understanding of infinity in a most unexpected way.

Cantor's key insight was that not all infinities are the same size.

This statement, when first encountered, tends to produce a moment of quiet disbelief. After all, infinity is already as large as one can imagine. How, then, can one infinite quantity be larger than another?

Cantor approached this question with admirable clarity. He considered sets of numbers and asked whether they could be placed in one-to-one correspondence with each other.

For example, the set of natural numbers (1, 2, 3, 4, ...) is infinite. The set of even numbers (2, 4, 6, 8, ...) is also infinite. At first glance, the set of even numbers appears to be smaller, since it consists of only every second number.

However, Cantor observed that one can pair each natural number with an even number:

$1 \leftrightarrow 2$

$2 \leftrightarrow 4$

$3 \leftrightarrow 6$

$4 \leftrightarrow 8$

and so on.

In this way, every natural number corresponds to exactly one even number, and vice versa. The two sets, despite appearances, have the same size.

This is a rather counterintuitive result.

It suggests that removing half the elements from an infinite set does not reduce its size. One might say that infinity is not particularly concerned with conservation.

Cantor then considered the set of real numbers—the numbers that include all possible decimal expansions. He showed that this set is larger than the set of natural numbers, in the sense that no one-to-one correspondence can be established between them.

This discovery introduced the idea of different "sizes" or cardinalities of infinity.

It also caused a certain amount of discomfort among mathematicians of the time, who were not entirely prepared for infinity to behave in such a complicated manner.

Cantor himself faced considerable opposition, though his ideas eventually gained acceptance and became a cornerstone of modern mathematics.

One cannot help but admire his persistence.

Infinity, it seems, rewards those who are willing to engage with it patiently, though it offers few concessions to common sense.

Beyond pure mathematics, infinity appears in many areas of science.

In physics, questions about the infinite arise in discussions of the universe itself. Is the universe infinite in extent, or does it have some form of boundary? Does time extend infinitely into the past and future, or does it have a beginning and an end?

These questions remain open, though they are approached with increasing sophistication.

In cosmology, infinity often appears not as a literal quantity but as a limiting concept—something that describes behaviour at extreme scales. Black holes, for example, involve singularities where certain physical quantities become infinite, at least according to current theories.

Such infinities are generally regarded with suspicion, as they suggest that our understanding is incomplete.

Mathematicians, by contrast, are rather more comfortable with infinity, provided it is handled properly.

In everyday life, infinity appears in more modest forms. Consider the decimal expansion of a number such as 1/3:

0.3333...

The threes continue indefinitely. This is an example of an infinite process contained within a finite representation.

Similarly, the number π has a decimal expansion that never ends and never repeats. It is, in a sense, infinitely detailed.

One could spend an entire lifetime calculating digits of π and still not reach the end, which is either inspiring or slightly exhausting, depending on one's disposition.

There is also a certain philosophical charm to infinity.

It invites us to consider the limits of human understanding, to acknowledge that some concepts extend beyond immediate intuition. It challenges the assumption that all quantities can be treated in the same way.

And yet, despite its peculiarities, infinity is not entirely beyond comprehension.

Through careful reasoning, mathematicians have developed tools that allow them to work with infinite processes in a precise and meaningful way. Limits, series, and set theory provide frameworks within which infinity can be understood, if not entirely tamed.

It is, perhaps, best thought of as a concept that resists domestication but responds well to gentle guidance.

There is also a certain humour in the way infinity behaves.

A hotel with infinitely many rooms, for example, can always accommodate more guests, even when it is already full. By moving each guest from room *n* to room *n+1*, one frees up room 1 for a new arrival.

This thought experiment, known as Hilbert's Hotel, illustrates the peculiar arithmetic of infinity. It is the sort of establishment one might prefer to avoid managing, though it provides excellent material for mathematical reflection.

In the end, infinity occupies a unique position in mathematics.

It is not a number in the ordinary sense, yet it is indispensable. It appears in calculations, in theories, and in the very structure of mathematics itself. It challenges intuition while enriching understanding.

Above all, it reminds us that mathematics is not limited to the familiar.

Beyond the comfortable territory of finite numbers lies a landscape that is vast, subtle, and occasionally bewildering. Infinity marks the boundary of this landscape—a concept that points beyond itself, inviting exploration.

It is, in its own way, a symbol of the limitless potential of mathematical thought.

And like many such symbols, it refuses to behave entirely as expected.

Which, one suspects, is precisely what makes it so interesting.

Bibliography – Chapter 5

Barrow, John D. *The Infinite Book*. Vintage, 2006.

Maor, Eli. *To Infinity and Beyond: A Cultural History of the Infinite.* Princeton University Press, 1987.

Cantor, Georg. *Contributions to the Founding of the Theory of Transfinite Numbers.* Dover Publications.

Stewart, Ian. *Concepts of Modern Mathematics.* Dover Publications.

Boyer, Carl B. *The History of the Calculus and Its Conceptual Development.* Dover Publications.

Kline, Morris. *Mathematics: The Loss of Certainty*. Oxford University Press.

SIX

CALCULUS AND THE ART OF CHANGE

There is something faintly unsettling about change.

One grows accustomed to things as they are—comfortably predictable, pleasantly stable—and then, quite without warning, they begin to alter. A moving car accelerates, a cup of tea cools, the stock market behaves in a manner best described as imaginative, and before one knows it, the world has shifted.

Mathematics, being a discipline that prefers order to surprise, eventually decided that this situation required attention.

Arithmetic, as we have seen, is perfectly adequate for dealing with static quantities. It handles fixed numbers with commendable efficiency. Geometry, too, describes shapes and spaces with admirable precision, provided they remain obligingly still.

But what of motion? What of growth, decay, and change itself?

For this, something more subtle was required.

And so, in the seventeenth century, mathematics acquired one of its most powerful tools: calculus.

It is generally agreed that calculus was developed independently by two individuals of considerable intellect and, one suspects, equally considerable determination: **Isaac Newton** in England and **Gottfried Wilhelm Leibniz** in Germany.

That both arrived at similar ideas independently is either a testament to the inevitability of calculus or a remarkable coincidence. That they subsequently engaged in a prolonged dispute over priority is, perhaps, less surprising.

Newton, a man not known for excessive modesty, referred to his method as the "method of fluxions." Leibniz, who possessed a rather more agreeable notation, introduced the symbols that are still used today.

History, in its usual fashion, has decided that both were correct and has quietly adopted Leibniz's notation, which is generally considered more civilised.

The essence of calculus lies in two closely related ideas: the study of rates of change and the accumulation of quantities.

These are known, respectively, as differentiation and integration.

Differentiation concerns itself with how things change. If one is travelling in a car, differentiation tells us how fast one is going at any given moment. It answers questions such as: "What is the rate of change of position with respect to time?"

Integration, on the other hand, deals with accumulation. Given a rate of change, it tells us how quantities build up over time. It answers questions such as: "How far have we travelled, given our speed over a period?"

Together, these two ideas form the backbone of calculus.

At first glance, this may not seem particularly extraordinary. After all, people had been observing motion and change for centuries. The novelty lies not in the questions, but in the method of answering them.

To see why this is necessary, consider a simple example.

Suppose one wishes to determine the speed of a moving object at a particular instant. If the object is travelling at a constant speed, this is straightforward enough. One measures the distance travelled over a given time and performs a modest calculation.

But what if the speed is changing?

In that case, the average speed over an interval does not fully describe the situation. One requires the instantaneous speed—the speed at a precise moment.

This presents a difficulty.

How does one measure something at an instant, when any measurement seems to require a span of time, however small?

It is here that calculus introduces a rather ingenious idea.

Instead of attempting to measure an instantaneous rate directly, one considers smaller and smaller intervals of time. As these intervals shrink, the average speed over the interval approaches the instantaneous speed.

This process of approaching a value is captured by the concept of a limit.

The derivative, which represents the rate of change, is defined as the limit of the average rate of change as the interval becomes infinitesimally small.

This may sound rather abstract, and indeed it is. Yet it provides a precise way of dealing with quantities that change continuously.

The elegance of the method lies in its ability to extract exact information from a process that appears inherently approximate.

If differentiation is about change, integration is about accumulation.

Imagine, for instance, that one knows the speed of a car at every moment during a journey. How does one determine the total distance travelled?

One approach is to divide the journey into small intervals, calculate the distance travelled in each interval, and then add them together. As the intervals become smaller, the approximation becomes more accurate.

In the limit, as the intervals approach zero, one obtains the exact distance.

This process is integration.

It is, in essence, a method for summing infinitely many infinitesimally small quantities.

There is something rather satisfying about this. Infinite processes, which we encountered in the previous chapter, are here harnessed in a practical and precise manner.

The connection between differentiation and integration is one of the most beautiful results in mathematics, known as the Fundamental Theorem of Calculus.

This theorem states, roughly speaking, that differentiation and integration are inverse processes. One undoes the other.

It is the sort of symmetry that mathematicians find deeply reassuring.

Having established the basic ideas, one might reasonably ask what calculus is actually used for.

The answer, in short, is almost everything.

In physics, calculus provides the language for describing motion, forces, and energy. Newton's laws of motion, which govern the behaviour of objects from falling apples to orbiting planets, are expressed in terms of derivatives.

Acceleration, for example, is the rate of change of velocity, which is itself the rate of change of position.

One begins to see that calculus is not merely useful in physics; it is essential.

In engineering, calculus is used to design structures, analyse systems, and optimise performance. Bridges, aircraft, and electrical circuits all rely on calculations that involve rates of change and accumulation.

In biology, calculus models population growth, the spread of diseases, and the dynamics of ecosystems. In economics, it describes how quantities such as cost, revenue, and demand change with various factors.

Even in everyday life, calculus makes quiet appearances. When a smartphone calculates the fastest route to a destination, or when a streaming service recommends content based on viewing patterns, calculus is often working behind the scenes.

It is, one might say, the unseen engine of modern technology.

Yet despite its widespread application, calculus retains a certain reputation for difficulty.

This is not entirely undeserved.

The concepts involved require a shift in thinking. One must become comfortable with limits, infinitesimals, and continuous change—ideas that do not always align with everyday intuition.

However, once this adjustment is made, calculus reveals a remarkable coherence.

Problems that initially appear complex can be approached systematically. Relationships that seem obscure become clear. Patterns emerge, guided by the principles of differentiation and integration.

There is also a certain aesthetic pleasure in the subject.

Equations in calculus often exhibit a kind of balance and symmetry. Solutions can be elegant, even graceful. A well-executed calculation has a satisfying completeness, rather like a neatly resolved piece of music.

It is perhaps this combination of utility and beauty that has secured calculus its central place in mathematics.

It is worth noting, however, that the development of calculus was not entirely without precedent.

Long before Newton and Leibniz, mathematicians in India had explored ideas that bear a striking resemblance to calculus. The Kerala school, particularly the work of **Madhava of Sangamagrama**, developed infinite series expansions for trigonometric functions.

These results anticipated aspects of calculus by several centuries.

The methods were not identical, and the formal framework of limits had not yet been established, but the underlying insight—that complex functions could be expressed as infinite sums—was already present.

It is a reminder that mathematical ideas often develop in multiple places, shaped by different contexts but converging toward similar conclusions.

Returning to calculus itself, one of its most powerful features is its ability to model continuous change.

In the natural world, many processes are continuous. Motion does not occur in discrete jumps; it flows smoothly. Temperature changes gradually. Growth and decay proceed over time.

Calculus provides a way to describe these processes with precision.

This is achieved through functions—mathematical expressions that relate one quantity to another. A function might describe how position depends on time, how temperature varies with location, or how population changes over years.

Differentiation and integration then allow us to analyse these functions.

We can determine where a function increases or decreases, where it reaches a maximum or minimum, and how it accumulates over an interval.

In this way, calculus transforms abstract relationships into concrete information.

It is, in effect, a tool for understanding how things behave.

There is also a certain philosophical dimension to calculus.

It deals with the interplay between the finite and the infinite, between the discrete and the continuous. It shows how infinite processes can produce finite results, and how local behaviour can determine global outcomes.

These ideas resonate beyond mathematics, touching on broader questions about change and continuity.

At the same time, calculus remains firmly grounded in practical application.

It is not a subject that exists solely for intellectual contemplation, though it certainly allows for that. It is a working tool, used daily by scientists, engineers, and analysts across a wide range of fields.

Its power lies in its generality.

Once one understands the basic principles, they can be applied to a vast array of problems. The same mathematical framework can describe the motion of planets, the flow of fluids, and the growth of populations.

This universality is one of the defining features of mathematics.

It is also one of the reasons calculus holds such a prominent place within it.

Of course, like any powerful tool, calculus requires careful handling.

Mistakes in calculation can lead to incorrect conclusions, and the abstract nature of the subject can sometimes obscure its meaning. It is therefore important to approach it with both precision and understanding.

Fortunately, with practice, the concepts become more intuitive.

What initially appears complex gradually becomes familiar. The symbols and notation, once mysterious, begin to make sense. The processes of differentiation and integration become tools that one can use with confidence.

There is, in this progression, a certain satisfaction.

It is the satisfaction of moving from confusion to clarity, from uncertainty to understanding.

In this respect, calculus is not merely a subject to be learned; it is an experience to be had.

One begins with a question about change and ends with a framework that explains it.

And along the way, one acquires a deeper appreciation for the structure of the world.

For the world, as it turns out, is in constant motion.

Objects move, quantities change, systems evolve. Calculus provides a way to describe and understand this dynamic behaviour.

It allows us to see not just where things are, but how they are changing.

And in doing so, it offers a perspective that is both practical and profound.

It reveals that change, far from being chaotic or arbitrary, often follows patterns that can be understood and predicted.

Which, when one considers it, is rather a comforting thought.

Even if the world insists on changing, it does so with a certain mathematical consistency.

And that, one feels, is something worth appreciating.

Bibliography – Chapter 6

Boyer, Carl B. *The History of the Calculus and Its Conceptual Development.* Dover Publications.

Stewart, James. *Calculus: Early Transcendentals.* Cengage Learning.

Newton, Isaac. *Philosophiæ Naturalis Principia Mathematica.*

Leibniz, Gottfried Wilhelm. Selected papers on calculus.

Joseph, George Gheverghese. *The Crest of the Peacock.* Princeton University Press.

Kline, Morris. *Calculus: An Intuitive and Physical Approach.* Dover Publications.

Struik, Dirk J. *A Concise History of Mathematics.* Dover Publications.

SEVEN

MATHEMATICS AND THE MUSIC OF THE UNIVERSE

There are few things in life that manage to be both entirely abstract and deeply moving. Music, rather curiously, is one of them. A sequence of vibrations in the air—nothing more, one might say—has the ability to stir emotions, evoke memories, and, on occasion, persuade perfectly sensible individuals to hum in public.

Mathematics, by contrast, is often accused of being rather less emotional. It deals in numbers, symbols, and logical relationships, none of which are generally associated with spontaneous displays of feeling.

And yet, the two are far more closely related than one might expect.

Indeed, one could argue—quietly, and perhaps over a cup of tea—that music is mathematics in a particularly agreeable disguise.

The connection between mathematics and music is not a recent discovery. It dates back to ancient Greece, where a certain fondness for both harmony and reasoning led to some rather intriguing observations.

The story is most often associated with **Pythagoras**, who is said to have noticed that musical harmony arises from simple numerical ratios. According to tradition, he observed that strings of different lengths produce harmonious sounds when their lengths are related by simple fractions.

A string half the length of another produces a note an octave higher. A ratio of 2:3 produces a pleasing interval known as a fifth. These relationships, once discovered, revealed that music was not merely an art but also a structure governed by number.

One suspects this must have been a deeply satisfying realisation.

For Pythagoras and his followers, this discovery suggested that the universe itself might be governed by similar principles. If harmony in music arises from numerical relationships, perhaps harmony in the cosmos does as well.

This idea became known as the "music of the spheres"—a poetic notion that the movements of celestial bodies produce a kind of cosmic harmony, inaudible to human ears but mathematically precise.

It is, admittedly, a rather charming idea.

Whether or not the planets are engaged in a silent symphony remains open to debate, but the underlying insight—that mathematics and harmony are closely linked—has proved enduring.

To understand this connection more concretely, it is helpful to consider what sound actually is.

Sound is a vibration—a wave that travels through a medium such as air. When an object vibrates, it creates fluctuations in pressure that propagate outward. These fluctuations, when they reach our ears, are interpreted as sound.

The properties of sound—pitch, volume, and timbre—are determined by the characteristics of these waves.

Pitch, for instance, depends on frequency—the number of vibrations per second. A higher frequency corresponds to a higher pitch. Volume depends on amplitude—the strength of the vibrations. Timbre, which gives each instrument its distinctive sound, depends on the combination of different frequencies present in the wave.

All of these properties can be described mathematically.

This is where mathematics begins to play a more explicit role in music.

When a musical instrument produces a note, it does not generate a single, pure frequency. Instead, it produces a fundamental frequency along with a series of overtones—higher frequencies that are integer multiples of the fundamental.

This collection of frequencies forms what is known as a harmonic series.

The harmonic series is, in many ways, the mathematical foundation of musical sound. It determines which combinations of notes sound harmonious and which do not.

When two notes are played together, their frequencies interact. If the ratio between the frequencies is simple—such as 2:1 or 3:2—the resulting sound is perceived as consonant, or pleasing. If the ratio is more complex, the sound may be dissonant, producing a sense of tension.

It is rather remarkable that our perception of harmony is so closely tied to numerical relationships.

One might say that the ear has a certain appreciation for mathematics.

Of course, music is not limited to simple ratios and harmonic series. As it developed, it became increasingly sophisticated, incorporating scales, rhythms, and structures that extend far beyond basic relationships.

Yet mathematics continues to provide the underlying framework.

Consider rhythm.

At its core, rhythm is a pattern in time. Notes are arranged in sequences, with durations that follow regular or irregular patterns. These patterns can be described using fractions and ratios, much like pitch.

A musical measure, for example, might be divided into equal parts—beats—which are then subdivided into smaller units. The relationships between these units create the rhythm of the piece.

Even complex rhythms, which may appear spontaneous or unpredictable, often follow underlying mathematical structures.

In this sense, music is not merely a sequence of sounds but a carefully organised pattern of events in time.

It is, quite literally, structured vibration.

As mathematics progressed, so too did its application to music.

In the eighteenth and nineteenth centuries, composers began to explore increasingly intricate musical forms. Counterpoint, harmony, and orchestration all became more elaborate, reflecting a growing understanding of musical structure.

At the same time, mathematicians were developing tools that would later prove invaluable in analysing sound.

One of the most important of these tools is the Fourier transform, developed by **Joseph Fourier**. This mathematical technique allows complex waveforms to be decomposed into their constituent frequencies.

In simpler terms, it reveals the individual notes hidden within a complex sound.

This has profound implications for music.

Using Fourier analysis, one can study the spectral content of musical sounds, understand the characteristics of different instruments, and even manipulate audio in precise ways.

Modern technologies such as digital music production, audio compression, and sound synthesis all rely on these mathematical principles.

It is, perhaps, one of the more practical demonstrations of how abstract mathematics finds its way into everyday life.

Yet beyond the technical aspects, there remains a deeper connection between mathematics and music—one that touches on aesthetics.

Mathematicians often speak of elegance, symmetry, and simplicity as qualities of a beautiful proof or equation. Musicians, too, value balance, proportion, and coherence.

A well-composed piece of music exhibits a kind of internal logic. Themes are introduced, developed, and resolved. Patterns are repeated and varied. There is a sense of structure that guides the listener through the piece.

This is not unlike the structure of a mathematical argument.

Both disciplines, in their own ways, seek to create order out of complexity.

There is also a shared appreciation for surprise.

In mathematics, a proof that arrives at an unexpected conclusion is often considered particularly elegant. In music, an unexpected chord or modulation can create a moment of delight.

These moments, though different in form, arise from the same underlying principle: the interplay between expectation and deviation.

One might say that both mathematics and music rely on a delicate balance between predictability and novelty.

Too much predictability, and the result becomes dull. Too much novelty, and it becomes incomprehensible.

The art lies in finding the right balance.

Interestingly, this balance can often be described mathematically.

Patterns that are too regular may lack interest, while those that are too irregular may lack coherence. The most engaging structures tend to exhibit a combination of order and variation.

This idea appears in many areas, from the rhythms of music to the patterns of nature.

Fractals, for example, exhibit self-similarity across different scales, creating structures that are both regular and complex. Similar principles can be found in musical compositions that repeat motifs while introducing variations.

The parallels are, at times, quite striking.

Even improvisation, which might seem the least structured aspect of music, often follows underlying rules. Jazz musicians, for instance, improvise within a framework of chords and scales, guided by an intuitive understanding of musical relationships.

It is, in effect, a spontaneous exploration of a mathematical space.

One might imagine that mathematics and music are engaged in a quiet conversation, each informing and enriching the other.

Of course, not all music is consciously mathematical, nor need it be. A listener does not typically analyse frequency ratios while enjoying a symphony, nor does a performer calculate Fourier transforms during a performance.

The mathematics operates in the background, providing structure without demanding attention.

This, perhaps, is part of its charm.

It is also worth noting that mathematics has influenced the way music is tuned.

The modern system of equal temperament, in which the octave is divided into twelve equal parts, is a compromise that allows instruments to play in multiple keys without retuning. This system is based on dividing the octave into equal logarithmic intervals, a decidedly mathematical procedure.

Earlier tuning systems, which relied more strictly on simple ratios, produced purer harmonies but were less flexible.

The choice of tuning system reflects a balance between mathematical precision and practical convenience.

As with many things, mathematics provides the framework, but human preference determines its application.

In recent years, the relationship between mathematics and music has expanded further, particularly with the advent of computers.

Algorithmic composition, for example, uses mathematical rules to generate music. Fractals, random processes, and other mathematical structures can be used to create compositions that are both novel and coherent.

Artificial intelligence, too, has begun to explore this territory, producing music that, while not yet entirely indistinguishable from human composition, is often surprisingly convincing.

It seems that mathematics, having quietly shaped music for centuries, is now taking a more active role.

And yet, despite all this, the essence of music remains unchanged.

It is still, at its heart, an arrangement of sounds that resonates with human experience.

Mathematics provides the structure, the framework, the underlying logic. But it is the human element—the interpretation, the expression, the emotion—that brings music to life.

This dual nature is what makes the relationship between mathematics and music so compelling.

One is precise, logical, and abstract. The other is expressive, intuitive, and deeply personal.

And yet, they are inextricably linked.

Perhaps this is not so surprising.

Both mathematics and music arise from the same human impulse: the desire to understand and to create, to find patterns and to express them.

In one, the patterns are symbols and equations. In the other, they are sounds and rhythms.

But the underlying principles are remarkably similar.

Both seek harmony.

Both value structure.

Both, in their own ways, reveal something about the nature of the universe.

And if, on occasion, the universe does indeed resemble a kind of grand composition—structured, dynamic, and occasionally surprising—then perhaps the old idea of the "music of the spheres" is not entirely fanciful after all.

At the very least, it provides a rather pleasant way of thinking about things.

And one suspects that, given the choice, the universe would not object to being described as musical.

After all, it has been keeping time rather well for quite some time now.

Bibliography – Chapter 7

Loy, Gareth. *Musimathics: The Mathematical Foundations of Music*. MIT Press.

Du Sautoy, Marcus. *The Music of the Primes*. HarperCollins.

Fourier, Joseph. *The Analytical Theory of Heat*. Cambridge University Press (translated editions).

Benson, Dave. *Music: A Mathematical Offering*. Cambridge University Press.

Hall, Rachel W. *Mathematics and Music: Composition, Perception, and Performance*. Springer.

Jeans, Sir James. *Science and Music*. Cambridge University Press.

EIGHT

Probability: When Mathematics Learns to Gamble

There is, in human affairs, a persistent and slightly uncomfortable truth: not everything can be predicted.

One may plan one's day with admirable precision, only to find that it is quietly derailed by traffic, weather, or a series of entirely avoidable decisions. One may invest with confidence and discover that the market has other ideas. One may even attempt to toss a coin and expect fairness, only to find that it has developed a curious preference for landing on the floor.

In short, the world contains uncertainty.

For a discipline such as mathematics, which has spent centuries cultivating certainty with almost obsessive care, this presents something of a challenge. Equations, after all, tend to produce definite answers. Two plus two is four, and remains so regardless of mood, weather, or economic policy.

Probability enters the scene at precisely this point.

It does not eliminate uncertainty—indeed, it acknowledges it—but it provides a way to understand and measure it. It is, one might say, mathematics' attempt to make peace with unpredictability.

The origins of probability are, perhaps inevitably, somewhat disreputable.

Unlike geometry, which arose from land measurement, or arithmetic, which emerged from counting, probability owes much of its early development to gambling. Games of chance, it turns out, have a remarkable ability to motivate mathematical inquiry, particularly when money is involved.

In the seventeenth century, a series of questions posed by gamblers led to a correspondence between two distinguished mathematicians: **Blaise Pascal** and **Pierre de Fermat**.

The problem, known as the "problem of points," concerned how to fairly divide the stakes of an unfinished game. If two players are engaged in a game of chance and are forced to stop before it concludes, how should the winnings be distributed based on their current positions?

It is the sort of question that, while perhaps not urgent in everyday life, becomes surprisingly important under certain circumstances.

Pascal and Fermat approached the problem by considering all possible outcomes of the remaining game and calculating the likelihood of each player winning. In doing so, they laid the foundations of probability theory.

It was a rather elegant solution to what might otherwise have resulted in a heated disagreement.

From these beginnings, probability developed into a systematic study of chance.

At its core lies a simple idea: the probability of an event is a measure of how likely it is to occur. This is often expressed as a number between 0 and 1, where 0 represents impossibility and 1 represents certainty.

A fair coin, for example, has a probability of 1/2 of landing heads. A standard die has a probability of 1/6 of landing on any given face.

These examples are straightforward, involving equally likely outcomes. But probability quickly extends beyond such simple cases.

Consider the weather.

One might hear that there is a 70% chance of rain tomorrow. This does not mean that it will rain on 70% of the day, nor that 70% of the city will experience rain. Rather, it reflects a model-based estimate of likelihood, derived from historical data and atmospheric conditions.

Probability, in this context, becomes a tool for making informed predictions.

It does not guarantee outcomes, but it provides guidance.

This distinction is important.

Probability does not tell us what will happen; it tells us what is likely to happen. The difference, while subtle, is significant. A highly probable event may fail to occur, and an unlikely event may, on occasion, take place.

This is the nature of chance.

One of the earliest insights in probability is the law of large numbers, which states that as the number of trials increases, the observed outcomes tend to approach the expected probabilities.

If one tosses a fair coin a few times, the results may appear erratic. But if one tosses it a thousand times, the proportion of heads and tails will likely be close to 50%.

This does not mean that the coin "remembers" previous outcomes or attempts to correct itself. Each toss remains independent. It is simply that, over many trials, patterns emerge.

This principle underlies much of statistical reasoning.

Closely related is the concept of expectation—the average outcome one can anticipate over many trials. In a game of chance, the expected value provides a measure of its fairness.

Casinos, it may be noted, are particularly attentive to this concept. Their games are carefully designed so that the expected value favours the house, albeit by a modest margin.

This ensures that, over time, the establishment remains comfortably solvent.

The gambler, by contrast, is often guided by intuition, optimism, and the occasional misunderstanding of probability.

One such misunderstanding is the gambler's fallacy—the belief that past outcomes influence future ones in independent events. After a series of heads in coin tosses, one might feel that a tail is "due."

Mathematically, this is not the case.

Each toss is independent, and the probability remains unchanged. The coin has no memory, no sense of fairness, and no interest in correcting perceived imbalances.

It simply continues to behave as a coin.

Probability, therefore, requires a certain discipline of thought. It demands that one distinguish between intuition and logic, between patterns that exist and patterns that are imagined.

This is not always easy.

Humans are, by nature, pattern-seeking creatures. We look for meaning in randomness, often finding it where none exists. Probability theory, in its quiet way, encourages a more cautious approach.

It reminds us that randomness can produce sequences that appear structured, and that apparent structure does not necessarily imply underlying order.

At the same time, probability reveals that randomness itself has structure.

This may seem contradictory, but it is not.

Random processes, when examined collectively, exhibit regularities. These regularities can be described mathematically, allowing us to understand and predict aggregate behaviour.

This is the basis of statistics.

Statistics takes the principles of probability and applies them to real-world data. It allows us to draw conclusions from samples, to estimate parameters, and to test hypotheses.

In medicine, statistics is used to evaluate treatments. In economics, it informs policy decisions. In science, it underpins experimental analysis.

Without probability, much of modern knowledge would lack a firm foundation.

One of the most elegant results in probability is Bayes' theorem, associated with **Thomas Bayes**.

Bayes' theorem provides a way to update probabilities based on new information. It allows us to revise our beliefs in light of evidence, a process that is both mathematically precise and philosophically appealing.

For example, if a medical test indicates a positive result, Bayes' theorem helps determine the probability that the patient actually has the condition, taking into account the accuracy of the test and the prevalence of the disease.

It is a powerful reminder that probability is not static; it evolves as new information becomes available.

In recent years, Bayesian methods have gained prominence in fields ranging from machine learning to decision theory. They offer a flexible framework for dealing with uncertainty in complex systems.

Returning to more familiar ground, probability also finds expression in everyday activities.

Weather forecasts, insurance policies, opinion polls, and even sports analytics all rely on probabilistic reasoning. These applications may differ in detail, but they share a common goal: to make sense of uncertainty.

There is, however, a certain irony in the use of probability.

While it provides tools for understanding randomness, it does not eliminate it. There will always be an element of unpredictability, a margin of uncertainty that resists complete control.

This is, perhaps, as it should be.

A world without uncertainty would be entirely predictable, and while this might simplify certain aspects of life, it would also render it rather dull.

Probability, then, occupies a curious position.

It acknowledges the limits of certainty while offering a means of navigating them. It bridges the gap between determinism and randomness, providing a framework for understanding both.

It is, in a sense, a philosophy as much as a mathematics.

One might even say that probability teaches a form of intellectual humility.

It reminds us that knowledge is often incomplete, that predictions are subject to uncertainty, and that outcomes are not always guaranteed.

At the same time, it offers reassurance.

Even in the presence of randomness, patterns exist. These patterns can be studied, understood, and, to some extent, anticipated.

This balance between uncertainty and structure is what makes probability so compelling.

It is also what makes it so widely applicable.

From the roll of a die to the behaviour of complex systems, probability provides a lens through which the unpredictable becomes, if not predictable, at least comprehensible.

And in doing so, it allows mathematics to engage with a world that is not always orderly.

A world where chance plays its part.

A world, in short, that occasionally behaves as though it were engaged in a game of its own.

Fortunately, mathematics has learned to play along.

And, on the whole, it plays rather well.

Bibliography - Chapter 8

Feller, William. *An Introduction to Probability Theory and Its Applications.* Wiley.

Mlodinow, Leonard. *The Drunkard's Walk: How Randomness Rules Our Lives.* Pantheon Books.

Salsburg, David. *The Lady Tasting Tea: How Statistics Revolutionized Science in the Twentieth Century.* W. H. Freeman.

Hacking, Ian. *The Emergence of Probability.* Cambridge University Press.

Bayes, Thomas. *An Essay towards Solving a Problem in the Doctrine of Chances.*

Kahneman, Daniel. *Thinking, Fast and Slow.* Farrar, Straus and Giroux.

NINE

Mathematics and the Architecture of Physics

There is something rather reassuring about a well-built structure.

A bridge that does not collapse, a building that remains upright despite the occasional enthusiasm of the weather, or even a modest chair that performs its duties without protest—all these inspire a quiet confidence. One may not understand the intricacies of their design, but one senses that they rest upon principles that are both sound and dependable.

Physics, it turns out, is much the same.

It is the study of how the universe is constructed, how its components interact, and how its grand machinery operates. And like any well-designed structure, physics relies upon a framework—an architecture—that gives it coherence and stability.

That framework is mathematics.

One might say that mathematics is the blueprint of the universe, while physics is the act of construction. The two are so closely intertwined that it is often difficult to tell where one ends and the other begins.

This relationship did not emerge overnight.

In the earliest stages of human inquiry, observations of nature were largely descriptive. The movement of the sun, the phases of the moon, the behaviour of falling objects—these were noted with care, but not necessarily explained with precision.

Gradually, however, patterns began to emerge.

The Greeks, with their characteristic fondness for order, sought to describe the natural world using geometry. The idea that celestial bodies move in perfect circles, for example, reflects a belief that nature adheres to geometric ideals.

While this particular notion turned out to be somewhat optimistic, it marked an important step: the attempt to describe physical phenomena using mathematical forms.

The true transformation occurred in the seventeenth century, when mathematics and physics entered into what can only be described as a remarkably productive partnership.

At the centre of this development was **Isaac Newton**, whose work unified motion, force, and mathematics in a manner that was both elegant and profoundly influential.

Newton's laws of motion provided a framework for understanding how objects move. More importantly, they expressed these principles in mathematical terms.

Force equals mass times acceleration—a statement that, while deceptively simple, encapsulates a vast range of physical behaviour.

Acceleration, as we have seen, is the rate of change of velocity, which itself is the rate of change of position. Thus, even in this basic formulation, calculus plays a central role.

With these tools, Newton was able to describe not only the motion of objects on Earth but also the movement of planets.

This was a remarkable achievement.

The same mathematical principles that govern a falling apple also govern the orbit of the moon. The universe, it seemed, was built according to a consistent set of rules.

There is something deeply satisfying about this unity.

It suggests that nature, for all its complexity, operates with a certain economy of principles. A small set of mathematical laws can describe a wide range of phenomena.

One might even say that the universe has a preference for efficiency.

As physics developed, mathematics continued to provide the language in which its laws were expressed.

In the study of electricity and magnetism, for example, the work of James Clerk Maxwell brought together a series of observations into a set of elegant equations. These equations describe how electric and magnetic fields interact, propagate, and influence charged particles.

They also predict the existence of electromagnetic waves—light itself.

It is a curious and rather delightful fact that something as familiar as light can be described by a set of equations.

The relationship between mathematics and physics becomes even more intriguing when one considers that mathematical structures often precede physical discovery.

In some cases, equations are developed for purely theoretical reasons, only to be found later to describe real phenomena.

This was notably the case with the work of **Paul Dirac**, whose equations predicted the existence of antimatter before it was observed experimentally.

Dirac's approach was guided by a belief in mathematical beauty. He sought equations that were elegant and symmetrical, and in doing so, uncovered aspects of reality that had not yet been detected.

This suggests that mathematics is not merely a tool for describing physics, but a guide to discovering it.

It is as though the universe has been constructed according to principles that mathematics is particularly well-suited to reveal.

Of course, this raises an interesting question.

Why should the universe be so accommodating?

Why should mathematical structures, developed in the human mind, correspond so closely to physical reality?

This question, which we encountered in earlier chapters, remains one of the great mysteries of science.

For now, it is sufficient to note that the correspondence exists—and that it is remarkably effective.

As physics progressed into the twentieth century, the relationship with mathematics became even more intricate.

Classical mechanics, with its clear and intuitive descriptions of motion, gave way to theories that were less immediately accessible but no less precise.

One such theory is quantum mechanics.

At the scale of atoms and subatomic particles, the behaviour of matter and energy becomes decidedly peculiar. Particles do not have definite positions and velocities in the classical sense. Instead, they are described by probabilities and wave functions.

These wave functions, which encode the possible states of a system, are mathematical objects. Their evolution over time is governed by equations that, while complex, provide accurate predictions of experimental results.

One might say that, at this level, reality itself becomes mathematical.

The certainty of classical physics is replaced by a probabilistic framework, yet the underlying structure remains firmly mathematical.

It is a subtle shift—from determinism to probability—but one that is handled with characteristic precision.

Another major development of the twentieth century is the theory of relativity, introduced by **Albert Einstein.**

Einstein's work transformed our understanding of space, time, and gravity. In special relativity, he showed that measurements of space and time depend on the observer's state of motion. In general relativity, he described gravity as the curvature of spacetime.

These ideas are expressed using sophisticated mathematics, particularly differential geometry.

Space and time are no longer separate entities but part of a unified structure—spacetime—which can be curved by the presence of mass and energy.

The mathematics involved is not trivial, but the underlying concept is remarkably elegant.

Objects move along paths determined by the geometry of spacetime. Gravity is not a force in the traditional sense but a manifestation of curvature.

It is, in a sense, geometry on a cosmic scale.

Once again, we see mathematics providing the architecture within which physics operates.

It defines the structure, establishes the relationships, and ensures consistency.

Without mathematics, these theories would be difficult, if not impossible, to formulate.

As one moves further into modern physics, the role of mathematics becomes even more pronounced.

Theories such as quantum field theory, which describe the behaviour of fundamental particles and forces, rely on advanced mathematical frameworks. Concepts such as symmetry, topology, and complex analysis play central roles.

Symmetry, in particular, has emerged as a guiding principle.

Physical laws often exhibit symmetries—properties that remain unchanged under certain transformations. These symmetries are described using mathematical structures known as groups.

The connection between symmetry and conservation laws is one of the most profound results in physics. For example, the symmetry of a system under time translation leads to the conservation of energy.

This relationship, formalised by Emmy Noether, illustrates the deep interplay between mathematics and physical principles.

It suggests that the laws of physics are not arbitrary but arise from underlying mathematical symmetries.

One might say that the universe is, at heart, a rather symmetrical place.

Of course, not all aspects of physics are perfectly symmetrical. There are asymmetries and irregularities that give rise to the diversity of phenomena we observe.

Yet even these deviations can often be described mathematically.

This brings us to an important point.

Mathematics does not merely describe idealised situations; it also accommodates complexity.

Real-world systems are often messy, involving multiple variables and interactions. Yet with sufficient mathematical sophistication, these systems can be analysed and understood.

This is particularly evident in fields such as fluid dynamics, where the behaviour of liquids and gases is governed by equations that are both elegant and, at times, challenging to solve.

The Navier–Stokes equations, for example, describe the motion of fluids. They are central to understanding phenomena ranging from ocean currents to airflow over an aircraft wing.

Despite their importance, these equations are notoriously difficult to solve in general.

This highlights an interesting aspect of the relationship between mathematics and physics.

Mathematics provides the framework, but solving the resulting equations can be a formidable task.

In some cases, exact solutions are not possible, and one must rely on approximations or numerical methods.

This is where computation enters the picture.

Modern physics relies heavily on computational techniques to simulate complex systems. Computers allow us to explore scenarios that would be difficult to analyse analytically.

Yet even here, mathematics remains essential.

The algorithms used in computation are based on mathematical principles, and the interpretation of results requires a solid understanding of the underlying theory.

Thus, the partnership between mathematics and physics continues to evolve.

It is not a static relationship but a dynamic one, adapting to new challenges and discoveries.

As we look to the future, this partnership shows no sign of diminishing.

Questions about the nature of dark matter and dark energy, the unification of forces, and the fundamental structure of the universe all require mathematical insight.

Theories such as string theory propose that the basic constituents of reality are not point particles but one-dimensional strings, vibrating in higher-dimensional spaces.

The mathematics involved is highly sophisticated, involving concepts that stretch the imagination.

Whether these theories accurately describe reality remains to be seen, but they illustrate the extent to which mathematics shapes our understanding of the physical world.

It is, in many ways, the language in which the universe is written.

Returning to more familiar ground, it is worth noting that the influence of mathematics extends beyond theoretical physics.

It underpins technology, engineering, and many aspects of everyday life.

The design of electronic devices, the analysis of signals, the development of materials—all rely on mathematical models derived from physical principles.

Thus, the architecture of physics, built upon mathematics, supports much of modern civilisation.

It is a structure both vast and intricate, extending from the smallest particles to the largest cosmic scales.

And yet, despite its complexity, it rests upon a foundation of remarkably simple ideas.

Numbers, equations, relationships.

From these, an entire universe is described.

There is something quietly extraordinary about this.

One might spend a lifetime studying the details, exploring the intricacies, and uncovering new connections. And still, the fundamental relationship remains: mathematics provides the structure, and physics reveals how that structure manifests in reality.

It is a partnership of remarkable depth and enduring significance.

And like any well-designed structure, it invites admiration—not only for its utility but for its elegance.

One cannot help but feel that, if the universe were inclined to express an opinion, it might take a certain pride in its mathematical architecture.

After all, it has held together rather well so far.

Bibliography – Chapter 9

Newton, Isaac. *Philosophiæ Naturalis Principia Mathematica.*

Einstein, Albert. *Relativity: The Special and the General Theory.*

Feynman, Richard. *The Character of Physical Law.* MIT Press.

Penrose, Roger. *The Road to Reality.* Jonathan Cape.

Dirac, Paul. *The Principles of Quantum Mechanics.* Oxford University Press.

Greene, Brian. *The Elegant Universe.* W. W. Norton.

Kaku, Michio. *Hyperspace.* Oxford University Press.

TEN

Patterns of Life: Mathematics in Biology

Life, at first glance, appears to be the very opposite of mathematics.

It is untidy, exuberant, occasionally unpredictable, and prone to doing things that no respectable equation would attempt without considerable hesitation. Plants grow in unexpected directions, animals behave in ways that defy easy categorisation, and human beings, in particular, display a talent for complexity that borders on the theatrical.

And yet, beneath this apparent disorder, there lies a quiet and persistent structure.

Biology, it turns out, is not exempt from mathematics. Indeed, it is gently permeated by it. Patterns, ratios, growth processes, and interactions—these all follow principles that mathematics is particularly well-suited to describe.

One might say that life, for all its variety, has a certain fondness for numbers.

This is not immediately obvious.

A forest, for instance, does not announce its mathematical properties. It presents itself as a collection of trees, leaves, and the occasional bird engaged in what appears to be a rather private conversation. But look a little more closely, and patterns begin to emerge.

Leaves are arranged in spirals. Branches divide in regular ways. Flowers exhibit symmetries that are both striking and consistent.

It is as though nature, while appearing spontaneous, is quietly following a set of guidelines.

One of the most frequently cited examples of such patterns is the Fibonacci sequence.

This sequence, which begins with 1, 1, 2, 3, 5, 8, and continues by adding each pair of preceding numbers, appears in a surprising number of biological contexts. The arrangement of leaves around a stem, the spirals of a pinecone, the pattern of seeds in a sunflower—all often reflect Fibonacci numbers.

It would be tempting to conclude that plants have been studying mathematics in their spare time.

The reality, of course, is somewhat less deliberate. These patterns arise from processes of growth that favour efficiency. Arranging leaves or seeds in certain ways allows for optimal exposure to sunlight or efficient packing.

Mathematics, in this context, describes the outcome of natural selection rather than guiding it directly.

Still, the correspondence is striking.

It suggests that certain mathematical structures are particularly well suited to solving biological problems.

Symmetry provides another example.

Many organisms exhibit bilateral symmetry—left and right sides that mirror each other. Others display radial symmetry, as seen in starfish or certain flowers.

Symmetry, in biological terms, often relates to function. It can influence movement, balance, and interaction with the environment.

Mathematically, symmetry is a property that can be described and analysed with precision.

The study of symmetry leads naturally to group theory, a branch of mathematics that examines transformations that leave certain properties unchanged. While this may sound somewhat abstract, it has practical applications in understanding biological structures.

The patterns on animal coats, for instance—stripes, spots, and other markings—can often be explained using mathematical models.

In the mid-twentieth century, **Alan Turing**, better known for his work in computing, proposed a model for how such patterns might arise. His reaction-diffusion theory describes how chemical substances, interacting and spreading through tissue, can produce stable patterns.

This was a remarkable insight.

It showed that complex biological patterns could emerge from relatively simple mathematical rules.

One begins to see a recurring theme: complexity arising from simplicity.

This is a hallmark of mathematics in biology.

Another area where mathematics plays a significant role is in the study of population dynamics.

Populations of organisms—whether they be bacteria, animals, or humans—change over time. They grow, decline, and interact with other populations.

Understanding these changes requires a mathematical framework.

One of the simplest models is exponential growth.

If a population grows at a rate proportional to its size, it will increase exponentially. This leads to rapid growth, at least initially.

However, in reality, resources are limited. Populations cannot grow indefinitely. This leads to more refined models, such as logistic growth, which incorporates limits imposed by the environment.

These models, while simple, provide valuable insights.

They help explain phenomena such as population booms, declines, and the effects of environmental constraints.

More complex models consider interactions between species.

Predator-prey relationships, for example, can be described using systems of differential equations. The classic example involves rabbits and foxes. As the rabbit population increases, the fox population grows due to an abundance of food. As the fox population increases, the rabbit population declines. This, in turn, leads to a decline in fox numbers, and the cycle continues.

These oscillations can be modelled mathematically, revealing patterns that might otherwise be difficult to discern.

It is a delicate balance—one that mathematics captures with surprising clarity.

Mathematics also plays a crucial role in genetics.

The inheritance of traits follows patterns that can be described using probability. The work of Gregor Mendel, though not initially recognised for its significance, laid the foundation for this understanding.

Mendel's laws describe how traits are passed from parents to offspring, and they can be expressed in terms of ratios and probabilities.

Modern genetics has expanded this framework considerably.

DNA, the molecule that carries genetic information, can be analysed using mathematical tools. Sequences of nucleotides are studied using algorithms, revealing patterns that inform our understanding of heredity and evolution.

Bioinformatics, a field that combines biology and computation, relies heavily on mathematics.

It is used to analyse genetic data, model biological systems, and develop new medical treatments.

Thus, mathematics not only describes life but also contributes to its study and improvement.

Another fascinating area is the study of networks.

Biological systems often consist of interconnected components. Neurons in the brain form complex networks. Ecological systems involve networks of interactions between species. Even within a cell, networks of chemical reactions take place.

These networks can be analysed using graph theory, a branch of mathematics concerned with connections and relationships.

Understanding these networks provides insights into how biological systems function.

The human brain, for instance, consists of billions of neurons connected in intricate ways. While the full complexity of this system remains beyond our current understanding, mathematical models help illuminate aspects of its structure and function.

Similarly, ecological networks reveal how species depend on one another.

Disruptions to these networks—such as the removal of a key species—can have cascading effects. Mathematics helps us understand these dynamics and, ideally, manage them more effectively.

It is, one might say, a way of bringing a degree of order to systems that might otherwise appear overwhelming.

Mathematics also finds application in the study of evolution.

Evolution, driven by natural selection, involves changes in populations over time. These changes can be modelled using mathematical frameworks that consider factors such as mutation, selection, and genetic drift.

Game theory, originally developed in economics, has been applied to evolutionary biology. It examines how strategies evolve in populations, particularly when individuals interact.

For example, cooperation and competition can be analysed using game-theoretic models. These models help explain why certain behaviours persist and how they evolve.

It is a curious intersection of disciplines—biology borrowing tools from mathematics that were themselves inspired by human behaviour.

And yet, the results are often illuminating.

One begins to see that mathematics provides a unifying language, capable of describing systems across different domains.

Even at the cellular level, mathematics plays a role.

The processes that sustain life—metabolism, signalling, replication—can be described using chemical kinetics and differential equations.

These models help us understand how cells function, how diseases develop, and how treatments might be designed.

In medicine, mathematical models are used to study the spread of diseases.

Epidemiology, the study of how diseases spread through populations, relies heavily on probability and differential equations.

The recent global experiences with pandemics have highlighted the importance of such models. They help predict the course of an outbreak, evaluate interventions, and inform policy decisions.

Of course, models are simplifications of reality.

They capture essential features but cannot account for every detail. This is both a strength and a limitation.

A model that is too simple may overlook important factors. One that is too complex may become unwieldy.

The art lies in finding the right balance.

This is, perhaps, where mathematics and biology differ in temperament.

Mathematics prefers clarity and precision. Biology embraces complexity and variation.

Yet the two complement each other.

Mathematics provides the tools to analyse and understand, while biology provides the richness of phenomena to be explored.

Together, they form a partnership that continues to evolve.

It is also worth noting that mathematics in biology often reveals unexpected beauty.

Patterns in nature—spirals, symmetries, branching structures—are not merely functional; they are aesthetically pleasing.

The structure of a leaf, the arrangement of petals, the form of a shell—these all exhibit a harmony that is both visual and mathematical.

One might say that nature has an eye for design.

Of course, this design is not intentional in the human sense. It arises from processes of growth and selection.

Yet the result is often striking.

It reminds us that mathematics is not confined to abstract symbols. It is present in the world around us, shaping the forms and patterns of life.

There is, in this realisation, a certain quiet delight.

One begins to see connections where none were apparent before.

A flower becomes more than a flower; it becomes a manifestation of mathematical principles. A population becomes more than a collection of individuals; it becomes a dynamic system governed by equations.

This perspective does not diminish the wonder of life.

On the contrary, it enhances it.

It reveals an additional layer of structure, a deeper level of understanding.

And it suggests that, even in the apparent chaos of living systems, there is an underlying order.

Not a rigid, inflexible order, but a flexible, adaptive one.

An order that accommodates variation and change.

An order that is, in its own way, alive.

Mathematics, in this context, is not an external imposition but a natural companion.

It does not dictate how life should behave; it describes how it does behave.

And in doing so, it provides a framework for understanding one of the most complex and fascinating aspects of the universe.

Life itself.

Which, when one considers it, is rather an impressive achievement.

Particularly for a collection of molecules that decided, quite some time ago, to organise themselves in interesting ways.

One suspects that mathematics, had it been present at the time, would have approved.

Quietly, of course.

Bibliography – Chapter 10

Murray, James D. *Mathematical Biology: I. An Introduction*. Springer.

Stewart, Ian. *Life's Other Secret: The New Mathematics of the Living World.* Wiley.

Ball, Philip. *The Self-Made Tapestry: Pattern Formation in Nature.* Oxford University Press.

Maynard Smith, John. *Evolution and the Theory of Games.* Cambridge University Press.

Nowak, Martin A. *Evolutionary Dynamics: Exploring the Equations of Life.* Harvard University Press.

Alon, Uri. *An Introduction to Systems Biology: Design Principles of Biological Circuits.* Chapman & Hall.

Turing, Alan. *The Chemical Basis of Morphogenesis.* Philosophical Transactions of the Royal Society B, 1952.

Edelstein-Keshet, Leah. *Mathematical Models in Biology*. SIAM.

Otto, Sarah P. and Day, Troy. *A Biologist's Guide to Mathematical Modeling in Ecology and Evolution*. Princeton University Press.

Keener, James and Sneyd, James. *Mathematical Physiology*. Springer.

ELEVEN

COMPUTERS: WHEN MATHEMATICS LEARNED TO THINK QUICKLY

There was a time, not particularly long ago, when computation required effort.

One does not mean the sort of effort associated with modern inconveniences—waiting a few seconds for a webpage to load, for instance—but genuine, sustained labour. Calculations were performed by hand, or with the assistance of mechanical devices whose enthusiasm often exceeded their reliability.

Tables of logarithms were consulted with a seriousness usually reserved for legal documents. Errors were made, corrected, and occasionally discovered much later, often at inconvenient moments.

In short, mathematics, while perfectly capable of thinking, did not think quickly.

The arrival of computers changed this rather dramatically.

At first glance, a computer appears to be a machine—a collection of circuits, wires, and components that perform operations with admirable speed. This is, of course, true. But it is also somewhat misleading.

A computer is not merely a machine. It is, in a very real sense, a physical embodiment of mathematical ideas.

It is mathematics given motion.

To understand how this came about, one must begin with a question that is both simple and surprisingly profound: what does it mean to compute?

For centuries, computation was a human activity. The term “computer” originally referred not to a machine but to a person—often a rather diligent individual tasked with performing calculations.

These human computers followed procedures—step-by-step methods for arriving at a result. Whether calculating astronomical positions, engineering measurements, or financial accounts, the process was essentially the same: apply rules systematically.

This observation—that computation consists of following rules—proved to be crucial.

In the early twentieth century, mathematicians began to formalise this idea. They asked whether all possible computations could be described in terms of a finite set of rules.

Among those who explored this question was **Alan Turing**, whose work would prove to be of lasting significance.

Turing introduced the concept of an abstract machine—now known as the Turing machine—that could perform computations by reading and writing symbols on a tape according to a set of rules.

This machine was not intended to be built. It was a theoretical construct, designed to capture the essence of computation.

Remarkably, it succeeded.

The Turing machine provided a precise definition of what it means for a function to be computable. It showed that any computation that can be performed algorithmically—that is, by following a set of rules—can be carried out by such a machine.

This idea, though abstract, laid the foundation for modern computing.

It suggested that computation could, in principle, be automated.

Of course, turning this principle into practice required a certain amount of engineering.

Early computing devices were mechanical, involving gears and levers. Charles Babbage, a nineteenth-century mathematician of considerable ambition, designed machines intended to perform calculations automatically.

His Analytical Engine, though never fully built in his lifetime, contained many of the features of a modern computer: a processing unit, memory, and the ability to follow instructions.

Babbage’s work was, in many ways, ahead of its time.

It would take the development of electronic technology in the twentieth century to realise the full potential of automated computation.

The first electronic computers were large, complex, and not particularly inclined toward portability. They occupied entire rooms and required careful maintenance. Yet they demonstrated something extraordinary: calculations that would take humans hours or days could be performed in seconds.

Mathematics, it seemed, had acquired a remarkable new assistant.

At the heart of this development lies a key concept: the algorithm.

An algorithm is a set of instructions for solving a problem. It is, in essence, a recipe for computation. Given an input, the algorithm specifies a sequence of steps that produce an output.

Algorithms are not new. They have existed in various forms for centuries. The methods developed by **Al-Khwarizmi** for solving equations are early examples.

What computers provide is the ability to execute algorithms with speed and precision.

This combination—mathematical logic and mechanical execution—is what makes computing so powerful.

It allows us to tackle problems that would be impractical, if not impossible, to solve by hand.

Consider, for example, the task of sorting a large list of numbers.

For a human, this is a tedious process. For a computer, it is routine. Efficient sorting algorithms can organise vast amounts of data in a fraction of a second.

Or consider the simulation of physical systems.

Using mathematical models, computers can simulate the behaviour of weather systems, the dynamics of galaxies, or the flow of fluids. These simulations provide insights that would be difficult to obtain through observation alone.

In this way, computers extend the reach of mathematics.

They do not replace mathematical reasoning; they amplify it.

Of course, the relationship between mathematics and computing is not one-sided.

Just as mathematics enables computing, computing has influenced mathematics.

Problems that were once considered purely theoretical have gained practical significance. New areas of mathematics have emerged, driven by the needs of computation.

Complexity theory, for instance, studies the efficiency of algorithms. It asks questions such as: how long does it take to solve a given problem? How much memory is required?

These questions are not merely academic.

They determine what can be computed in practice.

Some problems, while solvable in principle, require such vast amounts of time or resources that they remain effectively beyond reach.

This leads to an interesting distinction between what is possible and what is feasible.

A problem may have a solution, but that solution may not be accessible within a reasonable timeframe.

This is a rather sobering thought.

It suggests that even with powerful computers, there are limits to what can be achieved.

Yet these limits are themselves a subject of study, and mathematics provides the tools to explore them.

Another area where mathematics and computing intersect is cryptography.

In an age where information is transmitted and stored electronically, ensuring its security is of paramount importance.

Cryptography relies on mathematical principles to encode information in such a way that it can be accessed only by those with the appropriate key.

Modern cryptographic systems often depend on number theory—the study of properties of integers. Problems such as factoring large numbers or computing discrete logarithms are computationally difficult, making them suitable for encryption.

Here, mathematics serves not only as a tool for understanding but as a safeguard for communication.

It is a rather practical application of an otherwise abstract field.

The influence of computing extends further still.

Artificial intelligence, a field that has seen significant development in recent years, relies heavily on mathematical models.

Machine learning algorithms, for example, use statistical methods to identify patterns in data. They can recognise images, process language, and make predictions based on previous information.

While these systems are often described as "intelligent," it is worth noting that their operation is fundamentally mathematical.

They do not think in the human sense.

They compute.

They apply algorithms to data, adjusting parameters and refining models to improve performance.

It is, in effect, a sophisticated form of pattern recognition.

And yet, the results can be remarkably effective.

This raises interesting questions about the nature of intelligence and the role of mathematics in modelling it.

Can mathematical systems truly replicate human thought? Or do they merely simulate certain aspects of it?

These questions remain open, though they are explored with increasing interest.

Returning to the more immediate concerns of computation, one of the most striking features of computers is their speed.

Operations that would take humans considerable time are performed almost instantaneously. This has led to a transformation in how we approach problems.

Tasks that were once impractical are now routine.

Large datasets can be analysed, complex models can be simulated, and calculations can be performed with a level of precision that was previously unattainable.

This has implications across a wide range of fields.

In science, it enables more detailed and accurate models. In engineering, it supports design and optimisation. In everyday life, it powers technologies that we rely upon, often without a second thought.

Yet, despite this speed, the underlying principles remain mathematical.

A computer does not “understand” in the way a human does. It follows instructions—precisely, rapidly, and without deviation.

This is both its strength and its limitation.

It can perform calculations with remarkable efficiency, but it requires clear and correct instructions. An error in the algorithm leads to an error in the result, no matter how fast the computation.

This brings us back to the importance of mathematical reasoning.

Computers execute algorithms, but it is humans who design them.

The clarity of thought required to develop effective algorithms is, at its core, mathematical.

It involves understanding the structure of a problem, identifying patterns, and devising methods to address them.

In this sense, computing does not diminish the role of mathematics; it enhances it.

It provides new tools, new challenges, and new opportunities for exploration.

There is also a certain elegance in the way simple operations combine to produce complex behaviour.

At the lowest level, computers operate using binary digits—zeros and ones. These simple elements, arranged and manipulated according to logical rules, give rise to the vast capabilities of modern computing.

It is a reminder that complexity often arises from simplicity.

A theme, by now, that should feel pleasantly familiar.

As we consider the broader implications of computing, it becomes clear that we are witnessing an ongoing evolution.

Computers continue to become more powerful, more efficient, and more integrated into daily life. New paradigms, such as quantum computing, promise to further expand the boundaries of what is possible.

Quantum computing, which uses principles of quantum mechanics, represents a departure from classical computation. It has the potential to solve certain problems much more efficiently, though it also introduces new complexities.

The mathematics involved is, unsurprisingly, sophisticated.

Yet the underlying idea remains consistent: computation is a process governed by mathematical principles.

Whether carried out by hand, by mechanical devices, or by advanced electronic systems, it relies on the same fundamental concepts.

This continuity is rather reassuring.

It suggests that, despite technological change, the core ideas of mathematics remain stable.

They provide a foundation upon which new developments can be built.

In reflecting on the relationship between mathematics and computing, one might be tempted to ask whether computers have, in some sense, made mathematics easier.

The answer is both yes and no.

Yes, in that calculations can be performed quickly and accurately. No, in that the underlying ideas still require understanding.

A computer can solve an equation, but it cannot explain why the solution is meaningful.

That task remains, as ever, with us.

And perhaps that is as it should be.

For mathematics is not merely about obtaining results. It is about understanding relationships, exploring structures, and uncovering patterns.

Computers assist in this endeavour, but they do not replace it.

They are, if one may put it this way, exceptionally capable assistants.

Assistants that never tire, rarely complain, and perform their duties with remarkable efficiency.

Which, when one considers it, is rather more than one can say for most of us.

And so, mathematics, having learned to think, has now learned to think quickly.

The results have been transformative.

Yet the essence of the subject remains unchanged.

It is still a pursuit of understanding.

Only now, it proceeds at a rather brisker pace.

Bibliography – Chapter 11

Turing, Alan. *On Computable Numbers, with an Application to the Entscheidungsproblem.*

Knuth, Donald. *The Art of Computer Programming.* Addison-Wesley.

Sipser, Michael. *Introduction to the Theory of Computation.* Cengage Learning.

Cormen, Thomas et al. *Introduction to Algorithms.* MIT Press.

Shannon, Claude. *A Mathematical Theory of Communication.* Bell System Technical Journal.

Peterson, Ivars. *The Jungles of Randomness: A Mathematical Safari.* Wiley.

Nielsen, Michael and Chuang, Isaac. *Quantum Computation and Quantum Information.* Cambridge University Press.

TWELVE

Srinivasa Ramanujan: The Poet of Numbers

There are mathematicians, and then there are phenomena.

Most mathematicians, quite sensibly, proceed in an orderly fashion. They are trained, examined, corrected, and eventually entrusted with the responsibility of producing new ideas, preferably with proofs attached. Their work builds upon established knowledge, advancing the subject in careful, deliberate steps.

And then, on rare occasions, someone appears who seems to have bypassed the entire process.

Someone for whom mathematics is not a discipline to be learned, but a language already known.

Srinivasa Ramanujan was such a person.

To describe him as unusual would be accurate, but insufficient. To describe him as brilliant would be true, but inadequate. He was, in the most literal sense, extraordinary—a mathematician whose intuition operated at a level that continues to astonish even today.

His story, however, begins in circumstances that would not, at first glance, suggest mathematical destiny.

Ramanujan was born in 1887 in Erode, in what is now Tamil Nadu, and grew up in Kumbakonam. His family was not wealthy, and his early education was unremarkable in most respects. There was no immediate indication that he would become one of the most celebrated mathematicians in history.

Except, of course, for one detail.

He developed an early and intense fascination with numbers.

This fascination was not of the casual variety. It was not a passing interest or a mild curiosity. It was, by all accounts, consuming.

As a schoolboy, Ramanujan encountered a book that would shape his mathematical life: *A Synopsis of Elementary Results in Pure Mathematics* by G. S. Carr. This was not a conventional textbook. It contained thousands of results—formulas, identities, theorems—presented with minimal explanation.

For most readers, it would have been a rather daunting catalogue.

For Ramanujan, it was an invitation.

He did not simply read the results; he explored them. He reconstructed proofs, extended ideas, and developed new results of his own. Working largely in isolation, without formal training or guidance, he built a body of mathematical knowledge that was both deep and unconventional.

This is perhaps the first point at which one begins to suspect that Ramanujan was operating on a rather different wavelength.

His approach to mathematics was intuitive.

Where others relied on established methods, Ramanujan seemed to perceive patterns directly. He would write down formulas—often without proof—that turned out to be correct, sometimes astonishingly so.

This habit, while impressive, did not always endear him to the mathematical establishment.

In the conventional world of mathematics, results are expected to be justified. Proof is not merely a formality; it is the foundation upon which the subject rests.

Ramanujan, however, appeared to treat proofs as optional.

He produced results with such speed and confidence that one might be forgiven for thinking he was simply noting down observations rather than deriving them.

This created a certain difficulty.

Without proofs, his work was difficult to evaluate. It was clear that he possessed remarkable insight, but the lack of formal justification made it challenging for others to fully appreciate or verify his contributions.

For a time, this limited his recognition.

Ramanujan struggled academically in subjects that did not interest him, which, unfortunately, included most subjects other than mathematics. He failed examinations, lost scholarships, and found himself in a precarious position.

Yet he continued his mathematical work.

One might say that he had little choice.

Mathematics, for Ramanujan, was not an activity; it was an inclination.

Eventually, he secured a clerical position at the Madras Port Trust. It was not a role that required advanced mathematics, but it provided him with a modest income and, more importantly, time.

Time to think.

Time to write.

Time to fill notebooks with results that would later be recognised as extraordinary.

It was during this period that Ramanujan decided to reach out to mathematicians in England.

This was, by any standard, a bold decision.

He wrote letters containing some of his results—without proofs, but with enough detail to demonstrate their originality. These letters were sent to several prominent mathematicians.

Most did not respond.

One can understand their hesitation. The letters contained unfamiliar notation, unconventional methods, and claims that were difficult to assess.

But one recipient did respond.

This was **G. H. Hardy**, a mathematician at Cambridge known for his work in number theory.

Hardy, upon reading Ramanujan's letter, experienced what he later described as a mixture of bewilderment and admiration. Some of the results were incorrect. Many were unfamiliar. And several were undeniably brilliant.

Hardy recognised that he was dealing with something exceptional.

He shared the letter with his colleague **J. E. Littlewood**, and together they examined the results.

Their conclusion was clear: Ramanujan was a mathematician of extraordinary ability.

Hardy later remarked that the only comparable genius he had encountered was that of Euler and Jacobi—names not invoked lightly in mathematical circles.

An invitation was extended.

Ramanujan was asked to come to Cambridge.

This marked the beginning of one of the most remarkable collaborations in the history of mathematics.

When Ramanujan arrived in England in 1914, he entered a very different world.

Cambridge was a centre of mathematical activity, with established traditions, formal methods, and a strong emphasis on rigour.

Ramanujan, by contrast, brought an intuitive, almost instinctive approach.

The collaboration between Ramanujan and Hardy was, in many ways, a meeting of opposites.

Hardy provided structure, guidance, and an insistence on proof. Ramanujan provided insight, creativity, and a stream of results that often seemed to emerge fully formed.

Together, they achieved a great deal.

Ramanujan's work during this period covered a wide range of topics, including number theory, infinite series, and continued fractions. He produced results that were both deep and unexpected.

One of his most famous contributions is in the study of partitions—the ways in which numbers can be expressed as sums of smaller numbers.

For example, the number 4 can be partitioned in several ways:

4

3 + 1

2 + 2

2 + 1 + 1

1 + 1 + 1 + 1

The number of such partitions grows rapidly with larger numbers.

Ramanujan, working with Hardy, developed formulas to estimate these quantities with remarkable accuracy. Their work laid the foundation for further developments in number theory and combinatorics.

Another area in which Ramanujan excelled was the study of infinite series.

He discovered numerous identities involving π, many of which converge with astonishing speed. These formulas are still used today in high-precision calculations.

It is one of the more charming aspects of mathematics that results discovered under such unusual circumstances continue to find practical application.

Ramanujan also introduced what are now known as "mock theta functions"—objects that puzzled mathematicians for decades before their significance was fully understood.

His notebooks, filled with results and conjectures, have continued to yield insights long after his death.

It is, perhaps, the ultimate compliment: that one's work remains relevant across generations.

Yet, despite his success, Ramanujan's time in England was not easy.

The climate, the culture, and the food were all unfamiliar. His health began to deteriorate, and he faced periods of illness that limited his ability to work.

There was also the strain of adapting to a new environment.

And yet, he continued to produce remarkable mathematics.

In 1918, he was elected a Fellow of the Royal Society—one of the highest honours in science. It was a recognition of his contributions and his extraordinary talent.

He also became a Fellow of Trinity College, Cambridge.

These achievements, while significant, were overshadowed by his declining health.

In 1919, Ramanujan returned to India.

He continued to work, producing results even in his final months. But his health did not recover, and he passed away in 1920 at the age of 32.

It is difficult not to reflect on what might have been.

A longer life would likely have produced even more extraordinary contributions. And yet, what he achieved in such a short time is already remarkable.

Ramanujan's legacy extends beyond his specific results.

He represents a different way of engaging with mathematics—one that emphasises intuition, creativity, and a direct perception of patterns.

This is not to suggest that formal methods are unnecessary. On the contrary, the work of Hardy and others in providing proofs and structure was essential in making Ramanujan's results accessible.

But Ramanujan's approach reminds us that mathematics is not solely a logical exercise.

It is also an imaginative one.

There is, in his work, a sense of discovery that feels almost artistic.

Indeed, Ramanujan himself spoke of his results as being inspired by divine insight. Whether one interprets this literally or metaphorically, it reflects the depth of his engagement with the subject.

Mathematics, for him, was not merely a profession.

It was a calling.

There is also a certain humanity in his story.

He was not a product of privilege or extensive training. He worked in isolation, faced difficulties, and struggled with circumstances that might have discouraged others.

Yet he persisted.

And in doing so, he produced work that continues to inspire.

For readers encountering mathematics in a more modest way, Ramanujan's story offers both encouragement and perspective.

Encouragement, in that it shows what is possible. Perspective, in that it reminds us that even the most extraordinary achievements begin with curiosity.

One need not aspire to Ramanujan's level of genius to appreciate mathematics.

It is sufficient to engage with the subject, to explore its patterns, and to enjoy its insights.

Ramanujan's life, in its brevity and intensity, stands as a testament to the power of curiosity.

He saw patterns where others did not, pursued ideas with remarkable dedication, and left behind a body of work that continues to resonate.

If mathematics is, as we have suggested, a language of the universe, then Ramanujan was one of its most fluent speakers.

And like any great poet, he revealed possibilities that others had not yet imagined.

Which, when one considers it, is a rather fitting description.

For he was, in every sense, a poet of numbers.

Bibliography – Chapter 12

Kanigel, Robert. *The Man Who Knew Infinity: A Life of Srinivasa Ramanujan*. Scribner.

Hardy, G. H. *Ramanujan: Twelve Lectures on Subjects Suggested by His Life and Work*. Cambridge University Press.

Berndt, Bruce C. *Ramanujan's Notebooks*. Springer.

Andrews, George E. and Berndt, Bruce C. *Ramanujan's Lost Notebook*. Springer.

Ramanujan, Srinivasa. *Collected Papers*. Cambridge University Press.

Dyson, Freeman. *Missed Opportunities*. Bulletin of the American Mathematical Society.

THIRTEEN

OTHER INDIAN LUMINARIES OF MATHEMATICS

It is a curious habit of history that it occasionally selects one figure to represent an entire tradition.

This is convenient, of course. It simplifies matters. It allows us to speak of "the age of Newton" or "the era of Einstein" without becoming entangled in the contributions of their contemporaries. But it also has the unfortunate effect of obscuring the richness of the intellectual landscape from which such figures emerge.

In the case of Indian mathematics, the name of **Srinivasa Ramanujan** has, quite rightly, received widespread recognition. Yet to focus solely on Ramanujan would be to overlook a tradition that stretches back millennia—a tradition marked by ingenuity, originality, and a quiet persistence that has shaped mathematics in profound ways.

This chapter, therefore, is an attempt to step back and survey that landscape.

It is not a comprehensive account—such an undertaking would require several volumes and a certain degree of stamina—but rather a guided exploration of some of the key figures whose contributions have left an enduring mark.

We begin, as one often does, at the beginning.

Or at least, at a point sufficiently early to be considered foundational.

Aryabhata: The Quiet Revolutionary

In the year 499 CE, a mathematician named **Aryabhata** composed a work known as the *Aryabhatiya*. It is a text of remarkable brevity and density, containing results in arithmetic, algebra, and astronomy.

Aryabhata's approach was both practical and innovative.

He introduced methods for solving equations, approximating square roots, and working with trigonometric functions. His approximation of π, given as 3.1416, is impressively accurate for its time.

More striking, perhaps, is his approach to astronomy.

Aryabhata proposed that the apparent motion of the stars is due to the rotation of the Earth—a suggestion that, while not universally accepted at the time, reflects a willingness to question established views.

Mathematics and astronomy, in his work, are closely intertwined.

One might say that Aryabhata saw numbers not merely as abstract entities, but as tools for understanding the cosmos.

There is a certain elegance in this perspective.

Brahmagupta: The Architect of Zero

If Aryabhata laid foundations, **Brahmagupta** built upon them with considerable skill.

Writing in the seventh century, Brahmagupta produced the *Brahmasphutasiddhanta*, a text that addressed both mathematical and astronomical topics.

His most enduring contribution lies in his treatment of zero.

While the concept of zero had appeared earlier as a placeholder, Brahmagupta was among the first to treat it as a number with defined properties. He provided rules for arithmetic involving zero—how it behaves in addition, subtraction, and multiplication.

This may seem straightforward to modern readers, but it represents a significant conceptual advance.

To treat "nothing" as something—an entity that can be manipulated mathematically—is no small achievement.

Brahmagupta also made contributions to algebra, including solutions to quadratic equations and work on what are now known as Diophantine equations.

His methods were systematic, his results precise.

One suspects he would have approved of modern algebra, though perhaps with a few suggestions for improvement.

Bhaskara II: The Poet of Calculation

Several centuries later, **Bhaskara II** emerged as one of the leading mathematical minds of medieval India.

His works, particularly the *Lilavati* and *Bijaganita*, combine technical insight with a certain literary flair.

The *Lilavati*, in particular, is notable for its presentation. Problems are often framed in poetic language, inviting the reader to engage not only with the solution but with the narrative.

It is a reminder that mathematics need not be devoid of charm.

Bhaskara's contributions include advances in algebra, arithmetic, and what might be described as proto-calculus. He explored concepts related to rates of change and instantaneous motion—ideas that would later be formalised in calculus.

There is, in his work, a sense of anticipation.

He is approaching ideas that will be developed more fully centuries later.

And doing so with considerable skill.

The Kerala School: A Quiet Prelude to Calculus

In the southern region of India, between the fourteenth and sixteenth centuries, a group of mathematicians developed ideas that bear a striking resemblance to calculus.

This group, often referred to as the Kerala school, was led by **Madhava of Sangamagrama**.

Madhava's work on infinite series is particularly noteworthy.

He derived series expansions for trigonometric functions such as sine and cosine, and for π itself. These series allow for increasingly accurate approximations by summing more terms.

It is a powerful idea.

One that would later become central to calculus.

The Kerala school continued this line of work, refining methods and extending results. Mathematicians such as Nilakantha Somayaji and Jyesthadeva contributed to this tradition, producing texts that demonstrate a high level of sophistication.

What is particularly striking is that these developments occurred independently of European mathematics.

They represent a parallel evolution of ideas—one that highlights the universality of mathematical thought.

Pingala and the Mathematics of Prosody

Not all mathematical contributions arise from what one might traditionally consider mathematical contexts.

In ancient India, the study of poetic metre led to the development of combinatorial ideas.

The scholar **Pingala**, writing around the 3rd century BCE, analysed patterns of long and short syllables in poetry.

In doing so, he effectively developed a binary system.

Long and short syllables can be represented as two states, much like the 1s and 0s of modern computing. Pingala also explored combinations of these patterns, leading to results that resemble what we now call Pascal's triangle.

It is a delightful example of how mathematics can emerge from unexpected places.

One might say that poetry and mathematics, often considered distant cousins, are occasionally more closely related than one might think.

Hemachandra and the Fibonacci Connection

Continuing in the realm of combinatorics, the Jain scholar **Hemachandra** studied patterns of syllables in poetic metre.

In the process, he encountered a sequence of numbers that corresponds to what is now known as the Fibonacci sequence.

This sequence, which appears in various natural phenomena, was thus described in India well before Fibonacci himself.

It is another example of parallel discovery.

And a reminder that mathematical ideas often emerge in multiple contexts.

Narayana Pandita: The Combinatorial Craftsman

In the fourteenth century, **Narayana Pandita** made contributions to combinatorics and number theory.

His work includes methods for counting combinations and permutations, as well as problems involving magic squares.

These topics, while perhaps less immediately dramatic than celestial mechanics or infinite series, are fundamental to mathematics.

They form the basis for understanding arrangements, probabilities, and structures.

Narayana's work reflects a careful and methodical approach.

It is, one might say, the mathematics of arrangement.

The Broader Tradition

What emerges from this survey is not a collection of isolated achievements, but a tradition.

A tradition characterised by several notable features.

First, there is a strong connection between mathematics and astronomy.

Many Indian mathematicians were also astronomers, and their work reflects this dual focus. Mathematical methods were developed to solve practical problems related to celestial observation.

Second, there is an emphasis on computation and algorithms.

Procedures for calculation were carefully developed and transmitted. These methods were often expressed in verse, making them easier to memorise.

It is a rather efficient system.

Third, there is a willingness to explore abstract ideas.

The concept of zero, the use of infinite series, and the study of algebraic equations all demonstrate a capacity for abstraction that is both sophisticated and forward-looking.

Finally, there is a certain continuity.

Ideas developed in one period are refined and extended in later periods. The tradition evolves, adapting to new contexts while retaining its core insights.

A Reflection

It would be easy to present this history as a sequence of achievements, a catalogue of contributions that collectively advance the subject.

But there is also a human element.

These mathematicians worked in different times, under different conditions, often without the institutional support that modern researchers enjoy. Their work was driven by curiosity, by practical needs, and by a desire to understand.

They wrote in verse, calculated by hand, and transmitted knowledge through teaching and manuscripts.

And yet, their ideas have endured.

They form part of the foundation upon which modern mathematics is built.

It is a quiet legacy.

One that does not always receive the attention it deserves, but remains deeply influential.

Concluding Thoughts

Mathematics, as we have seen throughout this book, is a collective endeavour.

It is shaped by contributions from many cultures and individuals. The story of Indian mathematics is an integral part of this broader narrative.

It reminds us that mathematical insight is not confined to a particular place or time.

It arises wherever curiosity meets structure.

And it persists, carried forward by those who are willing to engage with it.

If there is a common thread among the figures discussed in this chapter, it is this: a commitment to understanding.

Whether through geometry, algebra, combinatorics, or astronomy, they sought to make sense of the world using the tools available to them.

In doing so, they expanded those tools.

And left behind a body of work that continues to inform and inspire.

Which, when one considers it, is a rather admirable achievement.

Even by mathematical standards.

Bibliography – Chapter 13

Plofker, Kim. *Mathematics in India*. Princeton University Press.

Joseph, George Gheverghese. *The Crest of the Peacock*. Princeton University Press.

Datta, B. and Singh, A. N. *History of Hindu Mathematics*. Asia Publishing House.

Sarasvati Amma, T. A. *Geometry in Ancient and Medieval India*. Motilal Banarsidass.

Pingree, David. *Census of the Exact Sciences in Sanskrit*. American Philosophical Society.

Hayashi, Takao. *Indian Mathematics*.

Bhaskara II. *Lilavati* and *Bijaganita* (translated editions).

Brahmagupta. *Brahmasphutasiddhanta* (translated editions).

Aryabhata. *Aryabhatiya* (translated editions).

FOURTEEN

Elegance: Why Mathematicians Fall in Love with Equations

There are, broadly speaking, two ways of solving a problem.

The first is to approach it with determination, apply a series of methods of increasing complexity, and eventually arrive at a correct answer, though not without a certain amount of intellectual perspiration.

The second is to arrive at the same answer with such clarity and simplicity that one wonders why it was ever considered difficult.

Mathematicians, perhaps unsurprisingly, have a strong preference for the second approach.

They call it elegance.

Elegance, in mathematics, is not a superficial quality. It is not merely a matter of presentation or style. It reflects something deeper—a sense that a solution captures the essence of a problem with minimal effort and maximum insight.

An elegant proof does not merely demonstrate that something is true; it reveals why it must be true.

This distinction, while subtle, is significant.

One may verify a result through calculation, checking each step with care. But an elegant argument provides understanding. It illuminates the structure underlying the result, making it appear almost inevitable.

It is, one might say, the difference between arriving at a destination and understanding the landscape through which one has travelled.

The appreciation of elegance is, in many ways, an acquired taste.

To those encountering mathematics for the first time, simplicity may not be immediately apparent. What appears simple to an experienced mathematician may seem anything but to a beginner.

Yet, with familiarity, patterns emerge.

One begins to recognise when a solution is unnecessarily complicated, when a problem has been approached from an awkward angle, and when a more direct path is available.

And when that direct path is found, there is a certain satisfaction.

It is not unlike solving a puzzle and discovering that the solution was, all along, within reach.

To illustrate this idea, consider a simple example.

Suppose one wishes to find the sum of the first hundred natural numbers:

$1 + 2 + 3 + \dots + 100$

One could, of course, add them sequentially. It would take some time, but the result would eventually be obtained.

Alternatively, one might notice a pattern.

Pairing the first and last terms, we have:

$1 + 100 = 101$

$2 + 99 = 101$

$3 + 98 = 101$

and so on.

There are fifty such pairs, each summing to 101. Thus, the total is:

$50 \times 101 = 5050$

This method is both efficient and illuminating.

It transforms a laborious calculation into a simple observation.

The story is often attributed to the young **Carl Friedrich Gauss**, who reportedly devised this method as a schoolboy.

Whether or not the story is entirely accurate, it serves as a useful illustration of elegance in action.

Elegance often involves recognising patterns.

Mathematics is rich with patterns—symmetries, repetitions, and relationships that, once identified, simplify complex problems.

These patterns are not always obvious. They require a certain attentiveness, a willingness to look beyond the surface.

But when they are found, they provide a powerful means of understanding.

Symmetry, in particular, is a recurring theme.

An equation or structure that exhibits symmetry often possesses a kind of balance that makes it easier to analyse. Symmetry can reveal invariants—properties that remain unchanged under certain transformations.

These invariants provide anchors, points of stability within a problem.

They guide the reasoning, suggesting paths that might otherwise remain hidden.

Consider, for instance, the equation:

$x^2 - 1 = 0$

At first glance, it appears straightforward. But its structure reveals a symmetry:

$(x - 1)(x + 1) = 0$

The solutions, $x = 1$ and $x = -1$, reflect this symmetry.

More generally, recognising that expressions can be factored, simplified, or transformed is often the key to an elegant solution.

This ability to transform is central to mathematics.

Equations are not static objects; they can be manipulated, rearranged, and interpreted in different ways. Each transformation reveals new aspects of the problem.

Elegance lies in choosing the right transformation.

Another aspect of elegance is generality.

A solution that applies to a wide range of problems is often considered more elegant than one that addresses a specific case.

General solutions reveal underlying principles.

They show that a particular result is not an isolated fact, but part of a broader pattern.

For example, the formula for the sum of the first n natural numbers:

$n(n + 1)/2$

is more elegant than computing each sum individually.

It encapsulates a general relationship in a compact form.

This compactness is another hallmark of elegance.

Mathematicians often admire results that achieve much with little—equations that express complex ideas in a concise manner.

Consider the equation:

$E = mc^2$

Associated with **Albert Einstein**, it relates energy, mass, and the speed of light.

The equation is remarkably simple, yet its implications are profound.

It exemplifies the idea that simplicity and depth are not mutually exclusive.

Indeed, they often go hand in hand.

This brings us to an important point.

Elegance is not merely aesthetic; it is functional.

An elegant solution is often easier to understand, to remember, and to apply. It reduces complexity, making problems more tractable.

In this sense, elegance is not a luxury but a practical advantage.

It is also, however, a source of pleasure.

Mathematicians frequently speak of the beauty of a proof or the elegance of a result. This language may seem unusual in a discipline associated with logic and precision, but it reflects a genuine experience.

There is a sense of satisfaction in encountering a solution that is both simple and insightful.

It is a quiet pleasure, perhaps, but a real one.

Of course, not all mathematics is elegant.

Some problems require lengthy calculations, intricate arguments, and considerable effort. There is no guarantee that a simple solution exists.

Indeed, part of the challenge is determining whether such a solution can be found.

Yet even in these cases, the search for elegance persists.

Mathematicians often revisit problems, seeking more direct approaches, more insightful arguments. A proof that is initially complex may later be simplified, revealing a deeper structure.

This process is ongoing.

It reflects a broader tendency in mathematics: the refinement of ideas.

Results are not merely established; they are improved.

Methods are not merely applied; they are optimised.

There is a continual movement toward clarity and simplicity.

This movement is not always straightforward.

What appears simple at one stage may be revealed as complex at another, and vice versa. The perception of elegance can change with understanding.

What matters is the direction of inquiry.

The pursuit of elegance encourages a particular way of thinking.

It emphasises insight over brute force, understanding over memorisation, and structure over complexity.

It invites one to ask not only whether a solution is correct, but whether it is the best possible solution.

This is, in many ways, a philosophical stance.

It reflects a belief that problems have underlying structures that can be uncovered, that simplicity is indicative of truth, and that beauty is a guide to understanding.

These ideas are not universally accepted, but they are widely held among mathematicians.

They influence how problems are approached, how results are evaluated, and how the subject evolves.

There is also a certain humility in the pursuit of elegance.

It requires acknowledging that one's initial approach may not be optimal, that there may be a simpler way, and that discovering it may require patience and persistence.

It is a reminder that mathematics is not merely about arriving at answers, but about refining the path to those answers.

This refinement is, in itself, a form of creativity.

It involves rethinking problems, exploring alternative perspectives, and recognising connections that were not immediately apparent.

It is, in a sense, an artistic process.

One might even say that mathematics, at its most elegant, resembles art.

It combines structure with creativity, precision with imagination.

It produces results that are both functional and aesthetically pleasing.

Of course, the comparison has its limits.

Mathematics is grounded in logic, while art allows for greater interpretive freedom. But the underlying appreciation of form, balance, and simplicity is shared.

Returning to equations, it is worth noting that they are not merely tools.

They are expressions of relationships.

An equation encapsulates a connection between quantities, a statement about how they interact.

When an equation is elegant, it captures this relationship with clarity.

It reveals the underlying structure in a way that is both concise and meaningful.

This is why mathematicians fall in love with equations.

Not because they are fond of symbols for their own sake, but because those symbols represent ideas.

Ideas that, when expressed elegantly, provide insight into the nature of the problem.

There is also a certain inevitability in an elegant result.

It feels as though it could not have been otherwise.

Once understood, it appears obvious.

This is, perhaps, the highest compliment one can pay to a mathematical argument.

That it makes the result seem inevitable.

And yet, arriving at such arguments is rarely straightforward.

It requires effort, exploration, and, occasionally, a willingness to reconsider one's assumptions.

It is a process that combines discipline with creativity.

And it is, in its own way, deeply satisfying.

As we approach the end of this exploration, it is worth reflecting on the broader significance of elegance.

It is not confined to mathematics.

In science, engineering, and even everyday problem-solving, elegant solutions are valued. They simplify, clarify, and reveal.

They provide not only answers but understanding.

Mathematics, however, offers a particularly clear expression of this idea.

It provides a domain in which elegance can be pursued systematically, where solutions can be evaluated not only for correctness but for quality.

And in doing so, it offers a model for thinking.

A way of approaching problems that values insight, simplicity, and coherence.

This is, perhaps, one of the most enduring contributions of mathematics.

Not merely the results it produces, but the standards it sets.

Standards that encourage clarity of thought, precision of expression, and an appreciation for the beauty of well-constructed ideas.

One might say that elegance is mathematics at its most refined.

A quiet achievement.

A moment when complexity gives way to simplicity.

And understanding, at last, feels complete.

Which, when one encounters it, is rather difficult not to admire.

Even if one is not, strictly speaking, a mathematician.

Bibliography – Chapter 14

Hardy, G. H. *A Mathematician's Apology*. Cambridge University Press.

Stewart, Ian. *17 Equations That Changed the World*. Profile Books.

Lockhart, Paul. *A Mathematician's Lament*. Bellevue Literary Press.

Polya, George. *How to Solve It*. Princeton University Press.

Courant, Richard and Robbins, Herbert. *What Is Mathematics?* Oxford University Press.

Tao, Terence. *Solving Mathematical Problems: A Personal Perspective*. Oxford University Press.

Weyl, Hermann. *Symmetry*. Princeton University Press.

FIFTEEN

The Mind of Mathematics: Discovery, Invention, and Intuition

There is a question, deceptively simple, that has managed to occupy mathematicians, philosophers, and the occasional curious observer for a rather long time.

What, exactly, is mathematics?

Not in the sense of definitions—those are readily available and, if one is determined, can be memorised—but in a deeper sense. Is mathematics something we create, like a novel or a piece of music? Or is it something we discover, like a continent that was always there, quietly minding its own business until someone thought to look?

The distinction is not trivial.

If mathematics is invented, then it is a product of the human mind, shaped by our preferences, our limitations, and our occasional fondness for symmetry. If it is discovered, then it exists independently of us, and we are merely uncovering truths that were already present.

One might say that mathematicians, while generally united in their appreciation of the subject, are quietly divided on this point.

Some regard mathematics as a creation—a system of symbols and rules devised to describe patterns. Others view it as a discovery—a realm of truth that exists whether or not we choose to engage with it.

And a few, perhaps wisely, reserve judgement.

To explore this question, it is helpful to consider how mathematics is actually done.

There is, in the popular imagination, a tendency to view mathematics as a strictly logical process. One begins with axioms, applies rules of reasoning, and arrives at conclusions. The process is orderly, precise, and, one might imagine, entirely devoid of surprises.

This is, of course, how mathematics is presented in textbooks.

It is not, however, how it is experienced.

Mathematicians do not typically sit down and proceed in a straight line from problem to solution. They explore, they experiment, they make conjectures, and they occasionally take paths that lead nowhere in particular.

There is, in short, a great deal of thinking involved.

And not all of it is logical in the narrow sense.

Insight, intuition, and imagination play significant roles.

The French mathematician **Henri Poincaré** wrote extensively about the process of mathematical discovery. He described how solutions often appear suddenly, after a period of unconscious thought.

One works on a problem, sets it aside, and then—perhaps while walking or engaging in some unrelated activity—the solution presents itself.

This phenomenon is not uncommon.

It suggests that mathematical thinking operates on multiple levels, some of which are not immediately accessible to conscious reasoning.

The mind, it seems, continues to work even when we are not actively directing it.

This raises an interesting possibility.

If mathematical ideas can emerge in this way, does that imply that they exist independently of our conscious effort? Are we, in some sense, discovering something that was already there?

The Austrian logician **Kurt Gödel** believed so.

Gödel held a view often described as mathematical Platonism—the idea that mathematical objects exist in an abstract realm, independent of human thought. According to this view, numbers, sets, and structures are not invented; they are discovered.

Mathematicians, in this framework, are explorers.

They navigate a landscape of abstract entities, uncovering relationships and truths that exist regardless of whether anyone observes them.

It is a compelling idea.

It also has the advantage of explaining why mathematics is so effective in describing the physical world. If mathematical structures are real, then it is perhaps less surprising that the universe conforms to them.

Of course, not everyone agrees.

Some mathematicians take a more pragmatic view, regarding mathematics as a human construct—a system of symbols and rules that we have developed to describe patterns.

From this perspective, mathematics is not discovered but invented.

We define the rules, we choose the axioms, and we explore the consequences.

There is, in this view, a certain flexibility.

Different systems can be constructed, each with its own internal logic. Geometry, for example, need not be Euclidean. Alternative geometries can be defined, leading to different conclusions.

This suggests that mathematics is not a single, fixed entity, but a collection of possible systems.

The truth, as is often the case, may lie somewhere in between.

Mathematics may be both invented and discovered.

We invent the language—the symbols, the notation, the frameworks—but in doing so, we uncover structures that appear to have an existence of their own.

It is rather like mapping a landscape.

We choose how to represent it, but the features we describe are not of our making.

This interplay between invention and discovery is one of the most intriguing aspects of mathematics.

It also brings us to the role of intuition.

Intuition, in mathematics, is often described with a certain caution. It is not always reliable, and it must be checked against formal reasoning. Yet it is indispensable.

It guides exploration.

It suggests directions.

It provides a sense of what might be true, even before a proof is found.

Consider the experience of encountering a mathematical pattern.

One may notice a relationship, test it in a few cases, and begin to suspect that it holds generally. This suspicion is not a proof, but it is a starting point.

It motivates further investigation.

In this way, intuition and logic work together.

Intuition proposes; logic verifies.

The balance between the two is delicate.

Too much reliance on intuition can lead to errors. Too much emphasis on formalism can stifle creativity.

The art of mathematics lies in maintaining this balance.

There is also a certain aesthetic dimension to intuition.

Mathematicians often speak of results as being "beautiful" or "elegant." This is not mere sentimentality. It reflects a sense that certain patterns possess a coherence and simplicity that resonate with our understanding.

The English mathematician **G. H. Hardy** famously wrote about the beauty of mathematics, comparing it to art.

He argued that a mathematician, like a painter or poet, creates patterns.

And these patterns, when well-constructed, possess a certain inevitability.

This sense of inevitability is closely related to intuition.

An elegant result often feels right, even before it is fully understood. It fits into a larger framework, aligning with other known results.

This alignment provides a kind of confirmation.

It suggests that the result is not an isolated fact, but part of a coherent structure.

This brings us to another intriguing aspect of mathematical thinking: the experience of recognition.

Mathematicians sometimes describe discovering a result as recognising something familiar.

This may seem odd.

After all, how can one recognise something that one has never encountered before?

Yet the description is apt.

A new idea, when understood, often feels as though it was always true—as though one has merely uncovered it rather than created it.

This experience lends support to the view that mathematics is discovered.

It suggests that mathematical truths have a kind of permanence.

They do not change with time or circumstance. They are not subject to revision in the way that empirical knowledge is.

Two plus two has always been four, and will remain so.

Of course, this permanence applies within a given framework.

If we change the rules, the results may differ. But within a consistent system, mathematical truths are stable.

This stability is one of the reasons mathematics is so reliable.

It provides a foundation upon which other disciplines can build.

And it does so with a degree of certainty that is difficult to find elsewhere.

Yet, for all its certainty, mathematics remains a creative endeavour.

New ideas are developed, new connections are made, and new structures are explored.

This creativity is not arbitrary.

It is guided by intuition, by a sense of coherence, and by the constraints of logic.

But it is creativity nonetheless.

Mathematicians do not merely uncover what is already known; they extend the boundaries of understanding.

They ask new questions, propose new frameworks, and explore new possibilities.

In doing so, they contribute to the evolving landscape of mathematics.

There is also a human dimension to this process.

Mathematics is not produced in isolation from the world.

It is shaped by the individuals who engage with it—their perspectives, their experiences, and their approaches.

Different mathematicians may approach the same problem in different ways, guided by different intuitions.

This diversity enriches the subject.

It leads to multiple perspectives, alternative methods, and a broader understanding.

At the same time, mathematics exhibits a remarkable degree of universality.

Results obtained in one context often find application in another.

Ideas developed in one culture are understood and extended in others.

This universality suggests that mathematics transcends individual perspectives.

It points to a shared structure, a common framework that underlies different approaches.

Returning to the question with which we began, one might say that mathematics occupies a unique position.

It is both a creation of the human mind and a discovery of underlying truths.

It is shaped by our thinking, yet it reveals patterns that seem to exist independently.

It is logical, yet it relies on intuition.

It is abstract, yet it describes the world with remarkable accuracy.

This combination of qualities is what makes mathematics so distinctive.

It is not easily categorised.

It resists simple definitions.

And perhaps that is part of its appeal.

There is, in mathematics, a sense of exploration.

Not of physical space, but of conceptual space.

One navigates ideas, follows connections, and uncovers relationships.

The journey is guided by intuition, supported by logic, and enriched by creativity.

And along the way, one encounters moments of clarity.

Moments when a problem resolves, a pattern emerges, or a connection becomes apparent.

These moments are often brief.

But they are deeply satisfying.

They provide a glimpse into the structure of mathematics—a structure that is both intricate and coherent.

One might even say that they offer a glimpse into the mind of mathematics itself.

A mind that is not conscious, but structured.

Not intentional, but consistent.

A mind that, in some sense, we participate in when we engage with mathematical ideas.

This may sound rather philosophical, and perhaps it is.

But it reflects a genuine aspect of mathematical experience.

To do mathematics is to enter a realm of ideas that are both created and discovered.

It is to engage with patterns that are at once familiar and new.

And it is to participate in a process that has been unfolding for centuries.

A process that continues, as new ideas are explored and new connections are made.

One does not need to resolve the question of invention versus discovery to appreciate this.

It is sufficient to recognise that mathematics offers a unique way of understanding.

A way that combines logic and intuition, structure and creativity.

And that, when approached with curiosity, reveals a world that is both orderly and surprisingly rich.

Which, when one considers it, is rather a pleasant place to spend one's time.

Even if, on occasion, it insists on asking difficult questions.

Bibliography – Chapter 15

Poincaré, Henri. *Science and Method*. Dover Publications.

Hadamard, Jacques. *The Psychology of Invention in the Mathematical Field*. Princeton University Press.

Gödel, Kurt. *Collected Works*. Oxford University Press.

Hardy, G. H. *A Mathematician's Apology*. Cambridge University Press.

Penrose, Roger. *The Emperor's New Mind*. Oxford University Press.

Tao, Terence. *Solving Mathematical Problems: A Personal Perspective*. Oxford University Press.

Polya, George. *How to Solve It*. Princeton University Press.

SIXTEEN

Why the Universe Speaks Mathematics

There are certain questions that, once asked, refuse to leave.

They linger, quietly at first, then with increasing persistence, until one is compelled to consider them more seriously than originally intended. One such question, which has made several appearances throughout this book in increasingly insistent forms, is this:

Why does the universe speak mathematics?

It is not a trivial curiosity.

The effectiveness of mathematics in describing the physical world is, as we have seen, both remarkable and somewhat puzzling. Equations developed in the quiet solitude of human thought turn out to describe the motion of planets, the behaviour of particles, the growth of populations, and even the structure of space and time.

This is not merely convenient.

It is, in a certain sense, astonishing.

One might have expected the universe to be less accommodating—to resist neat description, to behave in ways that defy simple patterns. Yet the opposite appears to be the case.

The universe is, to a surprising degree, intelligible.

And the language in which this intelligibility is expressed is mathematics.

The physicist **Eugene Wigner** famously referred to this as "the unreasonable effectiveness of mathematics." The phrase captures both the admiration and the puzzlement that accompany this observation.

Why should mathematics, a product of human abstraction, correspond so closely to physical reality?

There are, broadly speaking, several ways of approaching this question.

None is entirely satisfactory.

All are worth considering.

Mathematics as a Human Lens

One possibility is that mathematics works because we have designed it to work.

According to this view, mathematics is a human construct—a system of symbols and rules developed to describe patterns. We observe the world, identify regularities, and create mathematical frameworks that capture those regularities.

In this sense, mathematics does not reveal the structure of the universe; it reflects our interpretation of it.

The correspondence between mathematics and physics is therefore not surprising. It is the result of careful adjustment.

We refine our mathematical tools until they align with observation.

This is a reasonable position.

It emphasises the role of human creativity and adaptability. It suggests that mathematics is, in effect, a language we have developed to describe the world.

Yet it leaves something unexplained.

Why should such a language be possible at all?

Why should the universe exhibit patterns that can be captured so precisely?

One might argue that we notice patterns because they are there.

And that brings us to the next perspective.

Mathematics as the Structure of Reality

An alternative view is that mathematics is not merely a descriptive tool, but the underlying structure of reality itself.

According to this perspective, the universe is inherently mathematical. Its laws are not expressed in mathematics; they are mathematics.

This idea has been explored by physicists such as **Max Tegmark**, who has proposed that the physical universe is, at its core, a mathematical structure.

In this framework, the correspondence between mathematics and physics is not surprising.

It is inevitable.

We are not imposing mathematics on the universe; we are uncovering it.

This view has a certain elegance.

It explains why mathematical theories often predict phenomena before they are observed. It accounts for the deep connections between different areas of physics, all of which can be expressed within a unified mathematical framework.

It also raises some rather profound questions.

If the universe is a mathematical structure, what does that imply about our place within it? Are we part of this structure? Observers within a system that is, in some sense, abstract?

The implications are intriguing.

They are also, perhaps, a little unsettling.

The Role of Symmetry

One of the most compelling reasons mathematics works so well in physics is the presence of symmetry.

Symmetry, in the mathematical sense, refers to invariance under certain transformations. A system is symmetrical if it remains unchanged when subjected to a specific operation—such as rotation, reflection, or translation.

Physical laws exhibit a remarkable degree of symmetry.

They do not depend on where you are in space or when you perform an experiment. The laws of physics are the same here as they are elsewhere, now as they were in the past.

This uniformity can be expressed mathematically.

Symmetry leads to conservation laws. The invariance of physical laws under time translation leads to the conservation of energy. Invariance under spatial translation leads to the conservation of momentum.

These relationships, formalised by Emmy Noether, reveal a deep connection between mathematics and physics.

They suggest that the structure of physical laws is governed by mathematical principles.

Symmetry, in this sense, is not merely a feature of mathematics.

It is a feature of reality.

The Predictive Power of Mathematics

Another aspect of this relationship is the predictive power of mathematics.

Mathematical theories often extend beyond known observations. They make predictions—sometimes quite precise ones—that can be tested experimentally.

This is, perhaps, one of the most striking features of physics.

Equations do not merely describe what has been observed; they anticipate what has not yet been seen.

The discovery of Neptune, for example, was guided by mathematical predictions based on deviations in the orbit of Uranus. Similarly, the existence of antimatter was predicted by equations before it was observed.

These successes suggest that mathematics is not merely a descriptive tool.

It has a certain authority.

It tells us not only what is, but what must be, given certain conditions.

This is a powerful idea.

It implies that mathematical consistency can serve as a guide to physical truth.

The Simplicity Beneath Complexity

The universe, in its observable form, is extraordinarily complex.

It contains galaxies, stars, planets, ecosystems, and an assortment of phenomena that resist easy categorisation.

And yet, beneath this complexity, the underlying laws are remarkably simple.

Newton's laws, Maxwell's equations, Einstein's field equations—these are concise mathematical statements that describe a vast range of phenomena.

This simplicity is both surprising and reassuring.

It suggests that complexity arises from the interaction of simple rules.

Mathematics provides a way to express these rules.

And in doing so, it reveals a structure that might otherwise remain hidden.

One might say that mathematics allows us to see the framework beneath the detail.

The Human Mind and Mathematical Insight

There is, however, another dimension to this question.

Why are we able to understand mathematics at all?

If the universe is mathematical, and we are part of that universe, then our ability to comprehend mathematical structures may be less surprising.

Our minds, having evolved within this framework, may be attuned to its patterns.

This idea suggests a kind of resonance between the human mind and the structure of reality.

We are not external observers imposing order; we are participants in a system that exhibits order.

This does not fully resolve the question.

But it provides a perspective.

It suggests that the effectiveness of mathematics may be rooted in our position within the universe.

A Gentle Reflection

At this point, it may be tempting to seek a definitive answer.

To declare that mathematics is either invented or discovered, that the universe is either inherently mathematical or merely appears so.

Such conclusions, while satisfying, may be premature.

The relationship between mathematics and the universe is complex.

It involves elements of description, structure, perception, and perhaps something more subtle.

What can be said, with some confidence, is this:

Mathematics works.

It works with a consistency and precision that is difficult to ignore. It provides a framework for understanding the world, a language for expressing its laws, and a tool for exploring its behaviour.

Whether this effectiveness is "reasonable" or not is, perhaps, a matter of perspective.

From within the system, it may appear natural.

From outside, if such a viewpoint were possible, it might appear remarkable.

In either case, it invites reflection.

A Closing Thought

If one were inclined toward metaphor—and mathematics, despite its reputation, is not entirely averse to metaphor—one might say that the universe is written in a language.

A language that is precise, consistent, and, at times, surprisingly elegant.

We have, over centuries, learned to read parts of this language.

We have deciphered certain patterns, understood certain structures, and developed tools to explore further.

There remains much that is not yet understood.

But the progress thus far is encouraging.

It suggests that the universe is, in some sense, comprehensible.

And that mathematics is the key to that comprehension.

Why this should be so remains an open question.

One that continues to intrigue, to challenge, and to inspire.

Which, when one considers it, is rather fortunate.

For a question that has no answer is, at least, a question that invites continued exploration.

And mathematics, as we have seen, is particularly well suited to such endeavours.

Bibliography – Chapter 17

Wigner, Eugene. *The Unreasonable Effectiveness of Mathematics in the Natural Sciences.*

Tegmark, Max. *Our Mathematical Universe.* Knopf.

Penrose, Roger. *The Road to Reality.* Jonathan Cape.

Davies, Paul. *The Mind of God.* Simon & Schuster.

Einstein, Albert. *Ideas and Opinions.* Crown Publishers.

Barrow, John D. *The Constants of Nature.* Vintage.

Greene, Brian. *The Elegant Universe.* W. W. Norton.

SEVENTEEN

Does God Love Mathematics?

There are questions that belong comfortably within the domain of mathematics.

What is the value of a given expression? Does a particular equation have a solution? Can a structure be classified, a pattern generalised, a theorem proven?

These are, by and large, well-behaved questions. They have definitions, methods, and, with sufficient effort, answers.

And then there are questions of a rather different sort.

Questions that do not submit easily to calculation, that resist precise formulation, and that tend to linger somewhere between philosophy, intuition, and curiosity.

The present chapter concerns one such question:

Does God love mathematics?

It is, one must admit, not the sort of question that typically appears in examination papers. It is unlikely to yield to algebraic manipulation, and there is no standard method for deriving its answer.

Yet it arises naturally.

After all, if the universe exhibits such a remarkable degree of mathematical structure—if its laws are expressible in equations, if its patterns can be described with precision, if its behaviour aligns so closely with abstract reasoning—then one is inclined to wonder whether this is merely coincidental.

Or whether it reflects something deeper.

It is, perhaps, a question best approached with a certain degree of modesty.

And, if possible, a sense of humour.

The Mathematical Character of the Universe

Let us begin with an observation.

The universe, as we have seen, behaves in a manner that is strikingly mathematical. Physical laws are expressed in equations. Patterns in nature follow numerical relationships. Structures exhibit symmetry and order.

This is not a matter of convenience.

It is a pervasive feature of reality.

From the motion of planets to the structure of atoms, from the growth of organisms to the propagation of waves, mathematics provides an accurate and consistent description.

One might say that the universe appears to be organised according to mathematical principles.

This observation has not gone unnoticed.

The Italian scientist **Galileo Galilei** famously remarked that the book of nature is written in the language of mathematics.

It is a rather elegant way of putting the matter.

If nature is a book, mathematics is its script.

The question then becomes:

Who, if anyone, is the author?

Einstein and the Sense of Order

The physicist **Albert Einstein** approached this question with characteristic restraint.

He did not advocate a conventional theological view, but he did speak of a "cosmic religious feeling"—a sense of awe at the order and intelligibility of the universe.

For Einstein, the remarkable aspect of the universe was not merely that it exists, but that it can be understood.

That the human mind, with its limited capacity, is able to grasp the laws governing such a vast system.

This intelligibility, he suggested, is itself a mystery.

It is, in his words, "the most incomprehensible thing about the universe" that it is comprehensible.

This observation aligns closely with our theme.

If the universe is understandable through mathematics, then mathematics occupies a privileged position.

It is not merely a tool.

It is a key.

The Vedantic Perspective

Turning now to the Indian philosophical tradition, one finds a rather different, though intriguingly compatible, approach.

In Vedantic thought, the universe is not viewed as a collection of separate objects, but as an expression of a unified reality—Brahman.

This reality is described as infinite, unchanging, and underlying all phenomena.

The apparent diversity of the world is seen as a manifestation of this underlying unity.

One of the recurring themes in Vedanta is order.

The concept of *ṛta*—cosmic order—suggests that the universe operates according to principles that are consistent and intelligible.

This order is not imposed from outside; it is inherent.

Mathematics, in this context, can be seen as a way of perceiving that order.

It is not identical with Brahman, but it reflects aspects of the underlying structure.

There is also a notable affinity between the Vedantic notion of infinity and the mathematical concept.

The Upanishadic statement that the infinite remains unchanged when something is taken from it—"That is whole; this is whole"—resonates with the peculiar arithmetic of infinity we encountered earlier.

It suggests an intuitive understanding of properties that mathematics later formalised.

One might say that the philosophical and mathematical approaches, while distinct, converge on similar ideas.

Both recognise a form of underlying unity.

Both acknowledge the presence of structure.

And both, in their own ways, attempt to articulate it.

Indian Mathematicians and the Sense of Insight

The lives and works of Indian mathematicians also reflect a particular relationship with mathematics.

Srinivasa Ramanujan, as we have seen, often described his results as being inspired by divine insight. He spoke of mathematical formulas as expressions revealed to him, rather than constructed through deliberate reasoning.

This does not diminish the validity of his work.

On the contrary, it highlights a different mode of engagement.

Mathematics, for Ramanujan, was not merely a logical exercise.

It was an intuitive experience.

Similarly, earlier mathematicians such as **Bhaskara II** and **Aryabhata** approached mathematics with a sense of integration between knowledge and understanding.

Their work was not isolated from other domains.

It was connected to astronomy, philosophy, and, in some cases, spirituality.

This integrated perspective is worth noting.

It suggests that mathematics need not be confined to a purely technical framework.

It can be part of a broader inquiry into the nature of reality.

Is Mathematics Evidence of Design?

At this point, one might be tempted to draw a conclusion.

If the universe is mathematical, and if mathematics reflects order, then perhaps this order implies design.

And if there is design, perhaps there is a designer.

This line of reasoning is not new.

It has been considered, debated, and refined over centuries.

Yet it remains inconclusive.

There are alternative explanations.

One might argue that the universe simply possesses certain properties, and mathematics is the language we use to describe them. The apparent order may be a consequence of underlying physical laws, without implying intention.

Alternatively, one might suggest that only a universe with such order could give rise to observers capable of recognising it.

This is known as the anthropic principle.

It does not explain the origin of the order, but it offers a perspective on why we observe it.

Thus, the question remains open.

Mathematics may suggest order.

Whether that order implies design is a matter of interpretation.

Beauty as a Clue

There is, however, another aspect worth considering.

Mathematics is not only effective; it is often beautiful.

Equations, proofs, and structures can exhibit a kind of elegance that is difficult to describe but easy to recognise.

This aesthetic dimension is not incidental.

Mathematicians frequently use beauty as a guide.

A result that is simple, symmetrical, and coherent is often regarded as more likely to be true.

This is not a guarantee.

But it is a heuristic that has proved remarkably successful.

The question, then, is why mathematics should possess this aesthetic quality.

Why should the structures that describe reality also be pleasing to the mind?

One might argue that this is a coincidence.

Or one might suggest that it reflects a deeper alignment between the human mind and the structure of the universe.

Either way, it adds another layer to the question.

A Lightly Considered Possibility

At this stage, it may be appropriate to consider a modest proposal.

Not a conclusion, but a possibility.

If the universe exhibits mathematical order, and if this order is both intelligible and, at times, beautiful, then one might entertain the idea that mathematics is more than a descriptive tool.

That it reflects something fundamental.

Whether one chooses to interpret this in theological terms is a personal matter.

Some may see in it evidence of a guiding intelligence.

Others may regard it as an inherent property of reality.

Both perspectives have their merits.

And both, one suspects, would benefit from a certain degree of humility.

For the question, ultimately, may be beyond definitive resolution.

A Closing Reflection

It would be tempting to end with a definitive answer.

To declare, with confidence, whether or not God loves mathematics.

But such certainty would be out of place.

The question, after all, is not one that lends itself to proof.

What can be said is this:

Mathematics reveals a remarkable degree of order in the universe.

It provides a language through which that order can be understood.

It exhibits a beauty that resonates with human perception.

And it connects, in subtle ways, with philosophical traditions that emphasise unity and structure.

Whether this points to a divine preference for mathematics is, perhaps, less important than the fact that mathematics allows us to engage with these questions at all.

It offers a framework for exploration.

A way of thinking that is both precise and open-ended.

And in doing so, it invites us to consider not only how the universe works, but why it works in the way that it does.

Which, when one reflects upon it, is rather a remarkable invitation.

And one that mathematics, in its quiet and consistent manner, continues to extend.

Bibliography – Chapter 17

Einstein, Albert. *Ideas and Opinions.* Crown Publishers.

Davies, Paul. *The Mind of God.* Simon & Schuster.

Barrow, John D. *The Constants of Nature.* Vintage.

Jeans, James. *The Mysterious Universe.* Cambridge University Press.

Radhakrishnan, S. *The Principal Upanishads.* HarperCollins.

Easwaran, Eknath. *The Upanishads.* Nilgiri Press.

Joseph, George Gheverghese. *The Crest of the Peacock.* Princeton University Press.

Kanigel, Robert. *The Man Who Knew Infinity.* Scribner.

EIGHTEEN

THE FUTURE OF MATHEMATICS AND THE ENDLESS ADVENTURE

It is generally considered good form, when approaching the end of a book, to arrive at a conclusion.

A neat summary, perhaps. A sense of closure. The reassuring impression that matters have been resolved and that the reader may now proceed with a quiet feeling of completion.

Mathematics, however, is not particularly inclined toward such endings.

It has, over several millennia, developed a rather persistent habit of continuing.

And so, in approaching the future of mathematics, one must abandon any expectation of finality. There will be no concluding theorem, no ultimate equation that neatly encapsulates all that remains to be known.

Instead, there is something far more interesting.

An open horizon.

A Subject Without End

One of the most remarkable features of mathematics is that it is, in a very real sense, unfinished.

This is not due to a lack of effort. On the contrary, mathematicians have been working with considerable dedication for centuries. Entire libraries have been filled, theories developed, conjectures proposed and resolved.

And yet, for every question answered, several more appear.

This is not a sign of failure.

It is a sign of vitality.

Mathematics is not a static body of knowledge. It is an evolving landscape, shaped by inquiry, exploration, and the occasional unexpected discovery.

The deeper one goes, the more there is to see.

One might say that mathematics expands as it is explored.

New Frontiers

In the present age, mathematics finds itself at the centre of several rapidly developing fields.

Artificial intelligence, for instance, relies heavily on mathematical models. Algorithms learn from data, identify patterns, and make decisions based on statistical reasoning.

The underlying mathematics—probability, optimisation, linear algebra—is both sophisticated and essential.

Yet the questions extend beyond technique.

As machines become more capable, we are prompted to consider the nature of intelligence itself. What does it mean to "understand"? Can mathematical processes replicate human thought, or merely approximate certain aspects of it?

These are questions that lie at the intersection of mathematics, philosophy, and computer science.

They are, one suspects, unlikely to be resolved quickly.

Another frontier lies in the realm of quantum computing.

Unlike classical computers, which operate on binary digits, quantum computers utilise principles of quantum mechanics. They can, in theory, perform certain computations with remarkable efficiency.

The mathematics involved is intricate, involving complex numbers, linear operators, and probabilistic states.

It is, by any measure, a challenging field.

Yet its potential is considerable.

Problems that are currently intractable may become accessible. Cryptographic systems may need to be re-evaluated. New algorithms may emerge, reshaping our understanding of computation.

It is, in short, an area to watch with interest.

The Unresolved and the Unknown

Despite these advances, many fundamental questions remain unanswered.

In number theory, problems such as the Riemann Hypothesis continue to resist solution. This conjecture, concerning the distribution of prime numbers, has implications across mathematics.

It is one of several problems for which a solution would be both significant and, one imagines, warmly received.

In physics, the unification of general relativity and quantum mechanics remains an open challenge. The mathematical frameworks involved are well developed, but a complete synthesis has yet to be achieved.

In topology, geometry, and algebra, new structures continue to be explored. Connections between different areas of mathematics are discovered, revealing a web of relationships that grows ever more intricate.

This is, perhaps, one of the most appealing aspects of the subject.

There is always more to learn.

Mathematics and the Sciences

As we look to the future, the relationship between mathematics and the sciences is likely to deepen.

In biology, mathematical models are increasingly used to understand complex systems—genetic networks, ecosystems, and the dynamics of disease.

In economics, mathematical tools inform models of markets, behaviour, and decision-making.

In physics, mathematics continues to provide the framework for exploring the fundamental nature of reality.

These interactions are not one-sided.

Scientific developments often inspire new mathematics, just as mathematical insights enable scientific progress.

It is a partnership.

One that has proved remarkably productive.

The Human Element

Amid these developments, it is worth remembering that mathematics remains a human endeavour.

It is created, explored, and extended by individuals—each bringing their own perspective, intuition, and approach.

The tools may change. Computers may assist with calculation and simulation. Data may be processed at unprecedented scales.

But the core activity—the act of thinking, of recognising patterns, of making connections—remains unchanged.

Mathematics is not merely about results.

It is about understanding.

And understanding, for the moment at least, remains a distinctly human pursuit.

There is also a certain continuity.

The problems may become more complex, the methods more sophisticated, but the underlying curiosity persists.

The same questions that motivated early mathematicians—questions about number, space, pattern, and change—continue to inspire inquiry.

It is a thread that runs through the history of the subject.

And, one suspects, into its future.

A Return to Simplicity

There is a curious tendency in mathematics.

As ideas become more complex, there is often a parallel effort to simplify them.

Concepts are refined, frameworks are clarified, and results are expressed in more elegant forms.

This movement toward simplicity is not accidental.

It reflects a deeper principle: that understanding is often achieved not by adding complexity, but by reducing it.

The search for simplicity is, in many ways, the search for truth.

It is an ongoing process.

And it is one that will continue.

The Endless Adventure

At this point, one might be tempted to ask: where does it all lead?

What is the ultimate goal of mathematics?

The answer, perhaps, is that there is no single goal.

Mathematics is not a journey toward a fixed destination.

It is an exploration.

An ongoing adventure.

Each discovery opens new questions. Each insight reveals further possibilities. The landscape extends, inviting continued exploration.

This is not a deficiency.

It is the essence of the subject.

Mathematics thrives on inquiry.

It progresses through curiosity.

And it endures because there is always more to discover.

A Final Reflection

As we conclude this journey through mathematics, it is worth returning to a simple observation.

Mathematics began with counting.

With the recognition that quantities could be compared, that patterns could be observed, that relationships could be understood.

From these modest beginnings, it has grown into a vast and intricate discipline.

It has described the motion of planets, the structure of atoms, the patterns of life, and the behaviour of systems both simple and complex.

It has revealed connections between seemingly unrelated phenomena.

It has provided a language for understanding the universe.

And yet, at its heart, it remains accessible.

It begins with noticing.

With curiosity.

With the willingness to ask questions and to explore the answers.

This is, perhaps, the most important point.

Mathematics is not reserved for specialists.

It is a way of thinking.

A way of engaging with the world.

And, if approached with patience and curiosity, it offers a perspective that is both enlightening and, at times, quietly delightful.

A Parting Thought

If there is one idea that emerges from this book, it is this:

Mathematics is not finished.

It is not complete.

It is not, in any meaningful sense, closed.

It is an ongoing conversation.

One that has been taking place for centuries.

One that continues to evolve.

And one that, with a little inclination, anyone may join.

There is no requirement to solve great problems or to produce new theories.

It is sufficient to engage.

To notice patterns.

To appreciate structure.

To enjoy the elegance of a well-formed idea.

And perhaps, occasionally, to be surprised.

For mathematics, in all its precision, retains the capacity to surprise.

Which, when one considers it, is rather a pleasant quality.

Particularly in a subject that might otherwise be expected to behave predictably.

Bibliography – Chapter 18

Tao, Terence. *Solving Mathematical Problems: A Personal Perspective.* Oxford University Press.

Stewart, Ian. *The Great Mathematical Problems.* Profile Books.

Penrose, Roger. *The Road to Reality.* Jonathan Cape.

Du Sautoy, Marcus. *What We Cannot Know.* Fourth Estate.

Nielsen, Michael and Chuang, Isaac. *Quantum Computation and Quantum Information.* Cambridge University Press.

Goodfellow, Ian et al. *Deep Learning.* MIT Press.

Polya, George. *How to Solve It.* Princeton University Press.

Epilogue

It is customary, at the end of a journey, to look back.

Not in a sentimental manner—mathematics is not particularly fond of sentiment—but with a certain quiet acknowledgement of distance travelled. We have moved, over the course of these chapters, from counting small collections of objects to contemplating the structure of the universe itself.

Which, when one considers it, is a rather ambitious progression.

We began with numbers—modest, well-behaved entities that allowed early humans to keep track of sheep and, presumably, avoid awkward misunderstandings with neighbours. From there, mathematics expanded, acquiring geometry, algebra, and calculus, each extending our ability to describe and understand the world.

Along the way, matters became slightly more interesting.

Infinity appeared and refused to behave. Probability introduced uncertainty with an air of statistical respectability. Geometry abandoned flat surfaces and ventured into curved spacetime. Algebra persuaded numbers to adopt disguises and engage in conversation.

And then, rather unexpectedly, mathematics began to explain everything.

It described the motion of planets, the structure of matter, the patterns of life, and the rhythms of music. It entered biology, computing, and the architecture of physics with a quiet confidence, as though it had always intended to be there.

One might be forgiven for suspecting that mathematics had been quietly planning this all along.

Yet for all its reach, mathematics remains curiously personal.

It is, at its core, an activity of the mind.

It requires no elaborate equipment, no specialised environment. A pencil, a piece of paper, and a willingness to think are generally sufficient. Indeed, many significant developments have occurred under circumstances that were, at best, modest.

One is reminded that mathematics does not depend on external conditions.

It depends on attention.

This, perhaps, is its most democratic quality.

Anyone may engage with it.

Not everyone will pursue it to great depths, just as not everyone becomes a concert pianist or a professional athlete. But the door is open.

And stepping through it does not require permission.

Only curiosity.

There is also a certain reassurance in mathematics.

In a world that is often unpredictable, mathematics offers a degree of stability. Its truths do not change with fashion or opinion. A theorem, once proven, remains valid. A pattern, once recognised, does not suddenly alter itself out of convenience.

This constancy is comforting.

It provides a sense of order, a framework within which one can think clearly.

At the same time, mathematics is not rigid.

It evolves.

New ideas emerge, old ones are refined, and connections are discovered that were previously unimagined. It is both stable and dynamic—a combination that is, one might say, rather appealing.

And then there is the matter of beauty.

It would be remiss to conclude without acknowledging that mathematics, for all its precision, possesses an aesthetic dimension. This beauty is not always immediately apparent, particularly to those encountering the subject under less than ideal circumstances.

But it is there.

In the symmetry of an equation, the simplicity of a proof, the elegance of a solution—these are moments that mathematicians recognise and appreciate.

They are, in a quiet way, moments of delight.

Not the sort that provoke applause, but the sort that produce a small, satisfied pause.

A recognition that something fits.

That something makes sense.

That something, perhaps, is just right.

It is worth noting that this appreciation is not reserved for specialists.

One need not master advanced mathematics to experience it. Even simple ideas, when understood clearly, can provide a sense of satisfaction.

The realisation that a pattern exists, that a relationship holds, that a problem can be solved—these are accessible pleasures.

They require no special qualification.

Only attention.

And perhaps a little patience.

If this book has achieved anything, one hopes it has softened the boundary between mathematics and the reader.

Mathematics is often perceived as distant, difficult, or reserved for a select few. This perception, while understandable, is not entirely accurate.

Mathematics is, at heart, a human endeavour.

It is shaped by curiosity, guided by reasoning, and enriched by creativity. It is not beyond reach.

It simply requires a different way of looking.

A willingness to see patterns where there appeared to be none.

A readiness to ask questions, even if the answers are not immediately forthcoming.

And, occasionally, the courage to persist through confusion.

For confusion, it must be said, is not a sign of failure.

It is a sign that one is thinking.

And thinking, in mathematics as in life, is rarely a perfectly orderly process.

There are false starts, unexpected turns, and moments when the path forward is not immediately clear.

But there are also moments of clarity.

Moments when understanding arrives, often quietly, and with a certain inevitability.

These moments make the effort worthwhile.

They are, in a sense, the reward.

As we conclude, it may be appropriate to return to a question that has hovered, gently, over the latter chapters.

Why does mathematics work so well?

Why does it describe the universe with such precision?

Why does it appear, at times, almost as though it belongs there?

These questions remain open.

They invite reflection rather than resolution.

And perhaps that is as it should be.

For mathematics, like the universe it describes, is not entirely exhausted by our understanding.

There is always more to explore.

More to discover.

More to appreciate.

Which brings us, finally, to the reader.

If you have arrived at this point, you have travelled a considerable distance—not in miles, but in ideas. You have encountered concepts that may once have seemed unfamiliar, and perhaps found them less forbidding than expected.

If, along the way, mathematics has become slightly less intimidating, slightly more interesting, or even—dare one suggest it—occasionally enjoyable, then this book has served its purpose.

And if it has done a little more—if it has encouraged a moment of curiosity, a brief pause to notice a pattern, or a quiet appreciation of an elegant idea—then that is something to be valued.

For mathematics does not demand admiration.

It invites it.

Gently, persistently, and with a certain understated charm.

And if one accepts that invitation, one may find that mathematics is not merely a subject to be studied.

It is a companion.

A way of seeing.

A quiet, ongoing conversation with the structure of the world.

One that, rather conveniently, never quite comes to an end.

Which, when one reflects upon it, is probably for the best.

After all, it would be a shame if mathematics were to run out of things to say.

www.ingramcontent.com/pod-product-compliance
Lightning Source LLC
LaVergne TN
LVHW100324210926
848480LV00003B/1243